Environmental Plant Physiology

Environmental Plant Physiology

Botanical Strategies for a Climate Smart Planet

Vir Singh
Department of Environmental Science
G. B. Pant University of Agriculture and Technology, India

CRC Press
Taylor & Francis Group
Boca Raton London New York

CRC Press is an imprint of the
Taylor & Francis Group, an **informa** business

CRC Press
Taylor & Francis Group
6000 Broken Sound Parkway NW, Suite 300
Boca Raton, FL 33487-2742

© 2020 by Taylor & Francis Group, LLC
CRC Press is an imprint of Taylor & Francis Group, an Informa business

No claim to original U.S. Government works

Printed on acid-free paper

International Standard Book Number-13: 978-0-367-03042-1 (Hardback)

Visit the Taylor & Francis Web site at
http://www.taylorandfrancis.com

and the CRC Press Web site at
http://www.crcpress.com

Contents

Preface

The environment in our times is a matter of deep concern. No plan, no program, no project, and no socioeconomic development strategy can be carried out successfully if we ignore our environment. No sustainability can be imagined with the environment out of context. Not only our outstanding achievements but also our constant well-being and even our existence are all owing to the environment we live in. The environment is shared by innumerable species of plants, animals, and microorganisms. Each community of organisms occupying a specific ecosystem is comprised of organisms that are relatively more dependent on each other and share ecosystem resources in such a way that an ecological balance is always maintained. This ecological balance, in fact, is an attribute of environmental physiology.

Environment is not a static picture. It is held in dynamism. The dynamics of the environment in our times, however, is in doldrums. Environmental disruptions make up the common scenario of the globe. A state of ecological imbalance is a grim reality of our times. The dismal, gloomy, and hopeless state of our environment is thanks to indiscriminate anthropogenic interventions. What makes us absolutely nonplussed is the fact that human dimensions are so extensive and so intensive that the whole globe is infested with something that plagues the whole biosphere. Ecological integrity of the biosphere is breaking down rapidly. From a murky background of environmental disruptions and ecological imbalance emerges what is the biggest hue and cry of our world today: climate change. And this state of the biosphere is gradually and steadily leading to a catastrophic situation. This state of the biosphere, in fact, is also a natural outcome of environmental physiology—an alteration in plant physiological processes in response to changes in the normal climate pattern.

At this crucial juncture, when climate crises following unceasingly increasing greenhouse-gas concentrations are fast approaching a state of climate catastrophe, we earnestly need to evolve strategies by which we can hit the very roots of the climate crises. And this strategy is hidden in plant physiology. A deep and a thorough understanding of environmental plant physiology is of utmost necessity to formulate strategies for a climate-smart planet. *Environmental Plant Physiology* deals with the physiological processes of the plants as influenced by environmental factors reciprocally affecting the characteristic qualities of the environment and, thus, the very "behavior" of the climate. Environmental plant physiology is a relatively new branch of botany. The goal of the strategy we call the botanical strategy, of course, is to build up a climate-smart planet. A climate-smart planet will shower its benevolence on life. A climate-smart planet will be a healthy, vibrant, and sustainable Living Planet.

Environmental Plant Physiology: Botanical Strategies for a Climate Smart Planet offers a new application-oriented subject matter presented with a distinctive scientific flavor inviting students, teachers, scientists, planners, policy makers, and all those committed to the cause of planet Earth to study to enhance their knowledge base and evolve sound measures and tactics to build up a climate-smart planet, which is not only necessary but also an imperative.

The structure of the book reflects our goal of providing a scientifically integrative and multidisciplinary knowledge of vital plant physiological aspects including general aspects elaborating environment–plant–other organism relationships and the energy, nutrient, water, temperature, and allelochemical relations of the plants. High-altitude environments comprise little over 20% of the geographical area of Earth, but these ecosystems play critical roles in influencing climatic factors. Our strategies to deal with climate-change-related issues have to essentially incorporate the high-altitude ecosystems. The high-altitude habitats also offer many opportunities for cooling the earth, which have also been discussed. High-altitude physiology, therefore, has been discussed in a separate chapter. The penultimate chapter focuses on stress physiology. An understanding of various stresses on plant life caused by extreme environmental factors is essential to equip one with the knowledge of an invigorating dimension of plant physiology. The last chapter discusses the exhilarating matter of the physiological effects of climate change, a rare discussion hardly available in any text book of plant physiology published so far. Based on the current knowledge about climate change, an attempt has also been made in this chapter to synthesize a strategy to eventually genuinely realize the fact that a climate-smart planet is a sustainable planet.

Climate change is a cross-cutting issue. A number of disciplines deal, each in a unique way, with the changing climate. Environmental plant physiology is undoubtedly at the heart of all the disciplines. The present book will be stupendously useful for undergraduate, graduate, post-graduate, and research students and their teachers and researchers in various disciplines, such as plant sciences (including soil science, agriculture, and forestry), environmental science, ecology, ecological economics, climatology, and science of sustainability. Further, the book will also be of substantial use for chalking out strategies to effectively deal with climate change by environmental activists, policy makers, and climate workers in various government and nongovernment organizations working at local, national, regional, and international levels.

Acknowledgments

I am grateful to many professionals and their organizations who have granted permission to use some of their original matter (some figures and tables) in the book: Aruna T. Kumar, Indian Council of Agricultural Research (ICAR); Kuldeep Singh, director, National Bureau of Plant Genetic Resources (NBPGR); Chengcai Chu, Institute of Genetics and Developmental Biology, Chinese Academy of Sciences; editor-in-chief, *Journal of Emerging Technologies and Innovative Research (JETIR)*; Kevin Ward, NASA Earth Observatory Group/Earth Science Data Systems; PLS Clear, Oxford Publishing Limited; Barb Mattson, University of Maryland; and Monika Bright, University of Vienna, Austria.

I would like to express my gratitude to Prof. RD Gaur who taught me botany during my BSc Hons. degree program at Meerut University (now named Chaudhary Charan Singh University) and later supervised my research work leading to D. Phil. in botany (ecology) at HNB Garhwal University. I am indebted to the late Prof. Henryk Skolimowski of Michigan University for grafting a deep interest for ecology and ecophilosophy in my mind.

I wish to thank Renu Upadhyay and Jyotsna Jangra associated with CRC Press for their skillful guidance and for suggesting many ideas this book imbibes. My student Dr. Surindra Negi helped me by sketching a few figures. My occasional discussions with my students now engaged in professional tasks in India and abroad have led to the enrichment of the content of the book.

Finally, I wish to thank my wife, Gita; son, Pravesh; daughter, Silvi; and son-in-law, Nic Weber, for their moral support given throughout the project.

Vir Singh

Author

Dr. Vir Singh is professor of Environmental Science at G. B. Pant University of Agriculture & Technology. He has more than three decades' experience of teaching and research in forest ecology, environmental science, agroecology, animal sciences, environmental physiology, and natural resources management. Holding triple masters' (MSc botany, MSc Ag. animal nutrition, and MA sociology) and dual PhD degrees (botany with specialization in ecology and animal sciences), he has been educated and trained in many universities and institutes: Meerut University (now Chaudhary Charan Singh University), G. B. Pant University of Agriculture & Technology, HNB Garhwal University, Indira Gandhi National Open University (IGNOU), MP Bhoj (O) University, and Galilee College in Israel (now GIMI, Israel). He has been a Research Fellow at the International Centre for Integrated Mountain development (ICIMOD) based in Kathmandu and participated in courses in geoinformatics at Friedrich Schiller University (FSU) based in Jena, Germany. He has worked on many projects in collaboration with international institutes, including ICIMOD, ILRI, and the INNO-ASIA project sponsored by the German Federal Ministry BMBF. He has conducted several national and international conferences, symposia, and workshops. He has published several books, including recently in the limelight, *Fertilizing the Universe*, and more than 200 research articles and book chapters. Prof. Vir Singh is also a Climate Reality Leader committed to creating awareness about the ongoing climate change and its long-term implications on every walk of life and is also formulating programs and projects for climate-change mitigation.

1 Environment and Ecosystems
Physiological Basis of Ecology

Folks are like plants; we all lean toward the light.

Kris Carr

THE LUMENOSPHERE

What is the all-pervading factor in the universe that is also a key factor in the creation of life on Earth and which is the original source of all kinds of energy? It is light. This universe, in essence, is a home of light we can call lumenosphere (Latin: lumen = light; sphere, from Latin *sphaera* = globe, or range of a thing). It is not a thing that works. What works is a non-thing, that is, energy. It is the non-thing that makes the things work. All forms of energy in the universe emanate from light. The lumenosphere (or photosphere) accommodates countless galaxies, and each galaxy accommodates countless stars, and each star generates an infinite amount of radiation by means of nuclear reactions. And it is this radiation that illuminates the whole of its "home"—the lumenosphere—and keeps it going on.

The lumenosphere is not static in itself. Nothing, in fact, is static in the lumenosphere. The lumenosphere and everything in it is always in evolution. Evolution is a cosmic phenomenon that holds everything in the state of dynamism. Evolution, however, does not keep going on, on its own. It is triggered and propelled by the input of the energy of light. Things do not change on their own. There can be no change without energy. It is the energy that holds everything in a changing mode. The change is not random, abrupt, or directionless, or without a purpose. It has a rhythm, a definite direction, a hierarchical order, a purpose (Singh 2019). Such a change is what is called evolution. Evolution is a universal phenomenon. The lumenosphere is not just a physical dimension embracing all dimensions of existence. It is also a phenomenon in itself. Light is not just a form of energy comprising wave and particles (photons), it is a phenomenon, in fact, a phenomenon of all phenomena.

One of the phenomena of cosmic light is the biosphere. Powered by light, the biosphere itself is a living phenomenon of life. Reverberating with life, the biosphere is a living, perhaps the only living, oasis of the lumenosphere. Light is the only non-thing factor that has power to synthesize life within the biosphere—through a process known as photosynthesis. The other source of synthesizing life is chemical energy in some compounds, and the phenomenon associated with it is called chemosynthesis. Since the chemical energy is also rooted into the energy of light, chemosynthesis is also an indirect form of photosynthesis. It is through photosynthesis and chemosynthesis and energy in the organic molecules synthesized by these processes that the biosphere of the lumenosphere is replete with innumerable varieties of living organisms (have a glance at Figure 1.1). The diversity of life the biosphere accommodates is one of the most wonderful aspects of the lumenosphere. Our Living Planet and we all are part of this cosmic phenomenon. Environmental physiology helps us understand the phenomenon of life in the biosphere of Earth.

FIGURE 1.1 The cosmos as a lumenosphere. The biosphere of the earth—where solar energy flows in living organisms—is a "living oasis" of the lumenosphere.

THE BIOSPHERE

All ecosystems comprising a variety of communities of plants, animals, and microorganisms of the earth are integrated into a single stratum called *biosphere*. Biosphere (Greek: *bios* = life, *sphaira* = sphere) is the sphere or stratum of Earth's surface extending from a few kilometers into the atmosphere to a few kilometers into seas and oceans that supports and sustains life. The biosphere comprises biotic components (the living organisms) and the physical environment (abiotic components) that support life. The relationship between the biotic and abiotic components is reciprocal. Credit of the development of the term *biosphere* goes to English geologist Eduard Suess (1831–1914) and Russian physicist V. I. Vernadsky (1863–1945).

The biosphere includes parts of all the three components of the earth, viz., lithosphere, hydrosphere, and atmosphere and, in essence, is the sum of all the ecosystems of the earth. The biosphere is solar-powered and capable of converting solar energy into biochemical energy by means of a marvelous phenomenon called photosynthesis. Chlorophyll-containing plants, algae, and cyanobacteria are empowered to transform sunlight into biochemical energy, the energy of life. All animals, parasitic plants, fungi, and several unicellular organisms depend on photosynthesis directly or indirectly. There is yet a variety of microorganisms capable of transforming the energy of inorganic molecules into the energy of life through another spectacular phenomenon called chemosynthesis. There is a wonderful community around hydrothermal vents deep down in the oceans and seas exclusively dependent on chemosynthesis. The biosphere, thus, functions through photosynthesis and chemosynthesis. In other words, the biosphere is the only "home" to life in the universe that accommodates the phenomena of photosynthesis and chemosynthesis, the bases of all life on Earth.

In the biosphere, biotic and abiotic factors are in constant interaction: abiotic (inorganic) components become integral parts of the biotic (organic) components and vice versa. There is constantly a considerable exchange of matter between the two components. As a result, the living organisms do not just bear structures but are also a phenomenon in themselves, maintaining their structures through the exchange of matter with the physical environment. The physical environment includes the climatic factors that operate thermodynamics of the biosphere conducive to the phenomenon of life. The biosphere, stating differently, is a phenomenon in itself operationalizing the abiotic–biotic phenomenon of life.

The biosphere blossoms with enormous biodiversity. Prokaryotes (like bacteria), single-celled eukaryotes (like protozoa), fungi, plants, and animal species prospering within a variety of ecosystems of the biosphere, are estimated between 3 and 30 million, out of which some 1.4 million have been identified. Again, the intraspecies (genetic) diversity is also implausibly high. The biosphere is never in a static state. It is charged with the power of natural evolution. Hierarchical life, systems, and orders are all the spectacular outcomes of an unceasing evolution. Biodiversity, also a splendid feat of natural evolution, is organized into ecological groupings, such as populations of different species and communities. Biodiversity of life at every level comprises the richness of life and sum total of this richness and the factors and phenomena upholding, evolving, and sustaining the richness of life constitute the biosphere.

THE ORGANISM-ENVIRONMENT RELATIONSHIPS

An organism constitutes the smallest level of ecological hierarchy. It may be unicellular or multicellular in its structure but is an "autonomous" unit of the environment capable of performing its own functions. An organism, however, is not an independent entity. For its existence, sustenance and other essential activities, it depends on several other organisms directly and indirectly. All the organisms on planet Earth constitute a biotic component of the environment. They all are in continuous interaction with their physical environment (abiotic component of the environment). There is continuous exchange of materials between the organisms and their physical or abiotic environment. They derive energy, nutrients, water, and oxygen from the environment, which is indispensable for their maintenance, growth, and reproduction. They return to their environment whatever is over and above their maintenance, growth, and reproductive functions. These include the heat and solid, liquid, and gaseous wastes generated through metabolism.

The environment is an integrated whole. The abiotic and biotic components are woven into each other and are inseparable from each other. The quality of the physical environment has implications on the living organisms. And the living organisms have bearing on the quality of the physical environment. The two are in a reciprocal relationship. All the factors operating within the physical environment as well as all the living organisms in an environment are in interactions with each other. The study of the relationships/interactions among living organisms and between their abiotic and biotic components in the environment, called ecology, is an interesting subject to understand the overall processes of life, including dynamic equilibrium, environmental physiology, adaptation mechanisms, ecological evolution, community organization, ecosystems, biomes, and the nature of the biosphere.

The complex interactions between biotic and abiotic components (Table 1.1) in the environment determine the overall state of the environment. Organisms perform normally amidst appropriate environmental conditions set due to the abiotic–biotic interactions.

Biotic and abiotic components are held in dynamism. The matter alternates between abiotic and biotic components. For example, carbon gets incorporated into biomass through photosynthesis and becomes part of the biotic component. Nitrogen molecules get fixed into proteins and become part of the biotic component. Upon decomposition of the biomass, carbon (as CO_2) and nitrogen (N_2) again return to the abiotic component of the environment.

The interactions between abiotic and biotic components in an ecosystem determine which species can survive and sustain in a given ecosystem and also how the species have evolved in and adapted over time to that environment.

TABLE 1.1

Components of the Environment

Abiotic Components	Biotic Components
Climatic Factors	Green plants (photosynthesizers)
Radiation	Non-green plants
Heat	Symbionts
Temperature	Parasites
Atmospheric gases	Animals
Edaphic Factors	Human beings
Soil formation	Decomposers
Soil composition	—
Soil types	—
Soil profile	—
Physical properties of the soil	—

ENERGY AND NUTRIENT FLOWS THROUGH ECOSYSTEMS

A community occupying an ecosystem is self-reliant and self-sustainable. This is possible through nutrient and energy flows among organisms within an ecosystem. Ecosystem organisms can be categorized into producers, consumers, and decomposers.

Producers synthesize carbohydrates using abiotic resources, viz., atmospheric carbon dioxide and water, through photosynthesis and chemosynthesis. In photosynthesis, chlorophyll-containing plants, algae and blue-green algae (cyanobacteria), use the energy of light. In chemosynthesis, a variety of microorganisms use energy released from the oxidation of some inorganic molecules. The energy of light and inorganic molecules absorbed during photosynthesis and chemosynthesis, respectively, is incorporated in the chemical bonds of carbohydrates synthesized. Also referred to as autotrophs, the producers synthesize other organic compounds using the biochemical energy (transformed form of the energy in light and inorganic molecules) and absorb inorganic nutrients or minerals (another abiotic sources) from their habitats (from soil and water).

Consumers are dependent on the organic matter, other nutrients, and energy from other organisms. Herbivore consumers derive their foods (nutrients and energy) directly from the producers. Carnivore consumers derive their foods from the herbivores and/or other carnivores. Those that depend on herbivores are primary carnivores, and those dependent for their food on other carnivores are the carnivores of higher level, namely, secondary and tertiary or top carnivores.

Decomposers obtain nutrients and energy from dead and decaying organisms—producers as well as consumers. The decomposers break down the organic (biotic) compounds into their inorganic (abiotic) constituents. The abiotic constituents released from the decomposition of organic matter return to their respective pools: CO_2 and N_2 to the atmosphere and water to the hydrosphere. The mineral nutrients released through decomposition are recycled into soil for reuse by the producers. Energy released through catabolic processes and decomposition of organic compounds exits from the biosphere and returns to the space (Figure 1.2). The matter (the nutrients) flows in a cyclic manner, and the energy follows in a unidirectional flow—from organism to the space via atmosphere.

The energy-nutrient flows within an ecosystem, as described above, represent food chains that, in turn, are representative of the organism–environment relationships. Still better representatives are the complex combinations of these food chains called food webs. Transfer of energy and matter

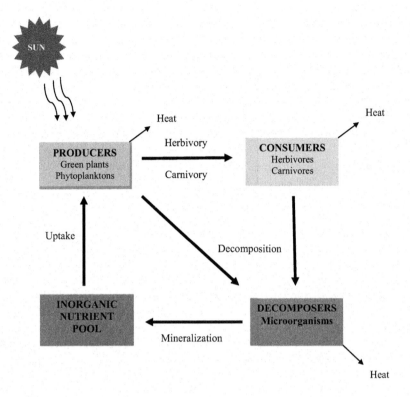

FIGURE 1.2 Energy flow through an ecosystem.

from one set of organisms (the tropic level) to the other involves loss of certain proportion of energy. We shall discuss organism–environment relationships by means of food chains and food webs in the following subheads.

FOOD CHAINS

A food chain refers to the transfer of food energy (along with nutrients) from the producers (the original sources) through a series of consumer organisms, from herbivores to top carnivores. Alternately, a food chain refers to the transfer of food energy from one trophic level (e.g., producers) to the next trophic levels (e.g., herbivores to top carnivores). All the ecosystems accommodate two types of food chains, viz. grazing food chains and detritus food chains.

Grazing food chains prevail in terrestrial and aquatic ecosystems. Photosynthesizing producers serve as the basis of these food chains. Solar radiation, therefore, is the primary source of energy in the grazing food chains. Among the consumers, often large animals, like tigers in the terrestrial ecosystems and sharks and whales in aquatic ecosystems, participate in these food chains as consumers. Some examples are as follows:

Terrestrial	Plant → Grasshopper → Small bird → Hawk
Ecosystem	Plant → Rabbit → Fox → Wolf → Tiger
	Plant → Grasshopper → Frog → Snake → Peacock
Aquatic	Phytoplankton → Zooplankton → Crustaceans → Insects → Small fish →
Ecosystem	Large fish
	Phytoplankton → Zooplankton → Small fish → Large fish → Shark
	Phytoplankton → Zooplankton → Small fish → Crane → Hawk

The detritus food chains prevail in the soil in terrestrial ecosystems and in the mud zone of aquatic ecosystems, like a pond. The primary source of energy is the organic matter of dead plants and animals, called detritus. However, as the organic matter emanates from photosynthesis, the detritus food chains also depend on solar energy indirectly. The primary consumers are the detritivores (detritus-eating organisms), like bacteria, protozoa, and fungi feeding on the dead organic matter saprophytically. The secondary consumers in the detritus food chains—insect larvae, crustaceans, mollusks, nematodes, etc.—feed on the primary consumers. These food chains are generally shorter than the grazing food chains. An example of the detritus food chain is:

Detritus → Protozoa → Insect larvae → Minnows → Large fish
Detritus → Fungi → Crustaceans → Small fish → Birds

Food chains, thus, are the nutritive relationships among biotic components of an ecosystem. In these relationships, a group of organisms feeds on smaller organisms and is eaten by larger organisms. The grazing food chains are generally longer than the detritus food chains. Food chains are linear. Omnivores and many carnivores occupy more than one trophic level in nature. Shorter food chains provide more of the available energy and vice versa. At each successive trophic level, 80%–90% of the energy is lost as heat, in accordance with the Second Law of Thermodynamics. The cost of respiration increases along successive trophic levels of a food chain. Respiration cost at the producers' level, on an average, is about 20%, at the herbivores' level 30%, and at the carnivores' level it is as high as 60%. The residual energy, in this way, goes on decreasing at each successive trophic level.

FOOD WEB

The food web includes an interconnection of several food chains operating in an ecosystem. A food chain is seldom independent. Linear arrangement of all the food chains in a biodiversity-laden ecosystem is hardly a reality. Several food chains create a complex network at various trophic levels formulating a food web and helping various organisms to participate in and get nourished from a number of feeding connections within a community of diverse organisms.

The interlinking of predatory, parasitic, and saprophytic chains provides alternative pathways for the organisms and enhances their food security. Food webs are of key ecological importance in the natural evolution of an ecosystem as they control population of the organisms with very high fecundity and determine the position of an organism in a biotic community. Examples of a terrestrial food web and an aquatic food web are illustrated in Figures 1.3 and 1.4, respectively.

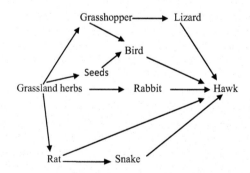

FIGURE 1.3 A terrestrial food web.

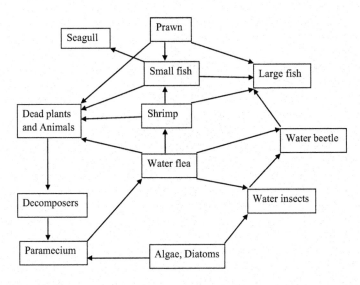

FIGURE 1.4 An aquatic food web.

ECOLOGICAL PYRAMIDS

The organism–environment relationship can also be described through graphical representation of some crucial ecological parameters, viz., number, biomass, and accumulated energy at various trophic levels in a food chain in an ecosystem, known as an ecological pyramid. The idea of ecological pyramids was devised by Charles Elton (1927). The shape of a pyramid is the characteristic of such a relationship. An ecological pyramid may be upright or straight, inverted, or spindle shaped, conveying the nature of the organism—environment relationship.

Ecological Pyramid of Numbers

Graphic representation of the number of individuals (or population size) of various species at different trophic levels in an ecosystem makes up an ecological pyramid of numbers.

In a grassland ecosystem or in a pond ecosystem, a predatory food chain assumes the shape of a straight or upright pyramid (Figure 1.5) wherein numbers of individuals go on decreasing successively from producers (green plants and phytoplankton in a grassland and a pond ecosystem, respectively) to herbivores to top carnivores.

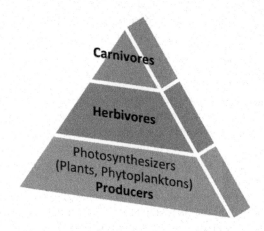

FIGURE 1.5 Ecological pyramid of numbers of a predatory food chain in a grassland or a pond ecosystem.

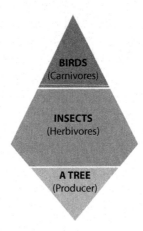

FIGURE 1.6   Spindle-shaped ecological pyramid of numbers.

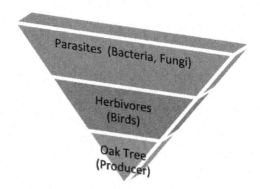

FIGURE 1.7   Ecological pyramid of numbers (oak-tree based).

In a tree-dominated ecosystem, a single tree (the producer) with an onslaught by numerous insects, which, in turn, are preyed upon by fewer birds, which can be preyed upon by the least number of hawks, serves as an ecosystem. Such a relationship among organisms, when represented graphically, assumes a spindle-shaped ecological pyramid (Figure 1.6).

When an oak tree harbors a large number of acorn-eating birds, which are further attacked by numerous parasites (like bacteria and fungi), the ecological pyramid formed is of an inverted shape (Figure 1.7).

Ecological Pyramid of Biomass

It is a graphic representation of biomass per unit area of different trophic levels in an ecosystem. In a terrestrial ecosystem, like a natural forest, the maximum amount will naturally be at the level of producers. The biomass will go on progressively decreasing from producers through herbivores to top carnivores. As a rule, only 10%–20% of the biomass is transferred through food chains from lower to higher trophic levels. Therefore, the ecological pyramid of biomass in a terrestrial ecosystem is straight or upright (Figure 1.8).

The ecological pyramid of biomass in an aquatic ecosystem, however, is an inverted one. The biomass of the producers (phytoplankton) is less than that of zooplankton, which is still less than that of top carnivores. An aquatic food chain, thus, represents an inverted (or spindle-shaped) pyramid (Figure 1.9).

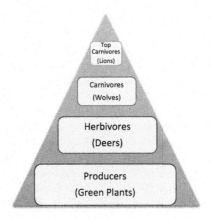

FIGURE 1.8   Pyramid of biomass in a terrestrial ecosystem.

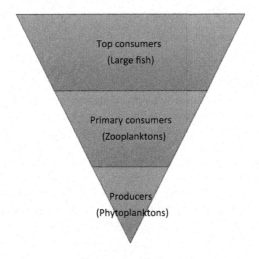

FIGURE 1.9   Pyramid of biomass in an aquatic ecosystem.

Ecological Pyramid of Energy

The pyramid of energy is a graphic representation of energy content trapped per unit area and time at different trophic levels of a food chain in an ecosystem. According to the Second Law of Thermodynamics, there is gradual decrease in energy content along successive trophic levels of a food chain prevailing in a community of organisms. At each transfer of energy through food, only 10%–20% is available to each successive trophic level. About 80%–90% of energy available at the lower trophic level is used in overcoming its entropy and in maintaining metabolic activities.

Retention of only 10% energy by the organisms at each successive trophic level is governed by the 10% law devised by Lindeman (1942). As the amount of energy retained by organisms goes on decreasing gradually along successive trophic levels, an ecological pyramid of energy is always vertical or upright (Figure 1.10).

If 15 calories are trapped by the grass in an ecosystem, only 0.0015 calorie is left for the top carnivores (Figure 1.8). Populations of the organisms living closest to producers are most sustainable. Sustainability of life goes on decreasing as the organisms exist farther from the producers. Life of the organisms farthest from the producers, that is, of top carnivores, is quite precarious.

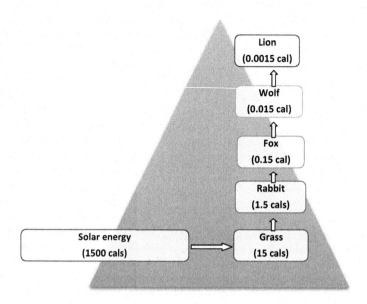

FIGURE 1.10 Pyramid of energy.

Since energy transfer from a set of organisms to another goes on becoming a limiting factor, there are only five trophic levels in nature (barring decomposers). Through its one-way travel of life (from producers to top carnivores), not much energy is left to support and sustain any more sets of living organisms.

THE SIXTH TROPHIC LEVEL

All the living organisms accommodating all the trophic levels in an ecosystem are eventually decomposed into individual components their structure is composed of. The decomposition process does not go waste. A category of the organisms, referred to as decomposers, derive energy from the dead and decaying organic matter of all the living organisms.

Decomposition is the reversal of photosynthesis and, as Singh (2019) suggests, may also be referred to as dephotosynthesis. Constituting three out of five kingdoms—prokaryotes, monerans, and fungi—the decomposers comprise the sixth trophic level of an ecosystem. This trophic level is not graphically represented as part of an ecological pyramid. The conventional ecologists generally count on just five trophic levels. The entire biomass of the biosphere eventually ends up into the sixth trophic level, which is not commonly taken into consideration. This trophic level may not obey the 10% law and comprises the largest proportion of biomass in the biosphere. The sixth trophic level, in fact, is the last phase of the nutrients from where they directly return to their original pool through what is called mineralization.

THE SEVENTH TROPHIC LEVEL

What is at the heart of the biosphere that is all-empowered to alter its structure, composition, and functioning? Who on Earth is capable of influencing even the very phenomenon of the biosphere, and even the evolutionary course of life of the Living Planet? It is undoubtedly the human species. The human species—*Homo sapiens*—has not only changed all the dimensions and phenomenon of life but has also emerged as a custodian of the Living Planet and even of the universe (Singh 2019).

The human species is the major consumer of the earth's natural resources. There is hardly any ecosystem on Earth that remains virgin. Human intervention in every ecosystem and in the life of each and every species is inevitable, indispensable! Human intervention in the planet's life ranges from the rare life on Earth's poles to the living chemosynthetic communities around hydrothermal vents 10,000 m deep into oceans.

What trophic level does the human species belong to? Producers? Yes, because almost all the earth's ecosystems are now anthropogenic ecosystems where the existence of the producers is designed by human beings. Human control is not only on species but also on their genotypes. Take the example of totally transformed ecosystems, that is, agricultural lands, in which not a single seed can germinate without the mercy of the human species. Humans have thus phenomenally changed even the phenomenon of life: photosynthesis.

Morphologically and anatomically the human species is herbivorous. We have tasted all the products plants of the earth produce. Knowledge of nutritive values, medicinal values, and all the hidden traits of numerous plants are at our finger tips. We are experts in cultivating all economic plants and utilizing the foods they produce.

The human species is genuinely an herbivore but has also developed carnivorous food habits. We cannot digest flesh. But we have developed food-processing technologies to digest animal products, consuming them as meat and other edible products. Not only do we relish the meat of herbivorous animals but also of carnivores of all trophic levels, and even of omnivores, like birds, fishes, and selected insects.

The human species consumes not only the organisms of all the five trophic levels but also of the sixth trophic level, that is, of the microorganisms. All brands of whisky, wine, beer, all other alcohol drinks, vinegar, and all fermented foods are prepared using a variety of microorganisms. The fermented drinks are, in fact, like the elixir of microorganisms.

Since the human species prevails in all the trophical systems and consumes most of the organisms inhabiting all the trophic levels in the ecosystems—terrestrial as well as aquatic—we cannot place humans in the so-far recognized trophics. Prevailing across all the trophics of the biosphere, the human species remains only a consumer, not a species consumed by other species—a supreme consumer, indeed! Singh (2019) suggests the human species to be categorized as the seventh trophic level of the biosphere. The all-pervading and all-consuming human species defies all the laws of the nature, including the 10% law of energy transfer. The processes of overconsumption of nature's resources, including the indiscriminate industrial processes managed by *Homo sapiens*, have influenced Earth's environment and ecological processes to the extent that even the climate of the biosphere is changing steadily, emerging as an unprecedented menace to the life of the planet.

WHAT IS ENVIRONMENTAL PHYSIOLOGY?

The entire biosphere operates by complex interactions of environmental factors, both biotic and abiotic. The environmental factors control and determine the functioning of the whole biosphere and, of course, of all the diverse communities of the diverse living organisms occupying a variety of ecosystems. Life on Earth, as a matter of fact, is an attribute of the specific environment of the planet. Living beings are made up of the matter of planet Earth and cosmic energy, that is, the energy of light. It is the energy of light and of inorganic molecules that synthesizes different molecules involving materials (e.g., carbon dioxide and water into glucose through photosynthesis or chemosynthesis). Existence, functioning, reproduction, and overall performance of the living organisms are all attributable to normal functioning of the organisms (the functioning of their cells, tissues, organs, and systems). We study the normal functions of the living organisms and/or of their parts under a biological discipline called physiology.

Physiology, in fact, pertains to the biochemical processes occurring within the internal environment of an organism—at the level of the cells, tissues, organs, and systems of the organism. The internal environment of an organism, however, is not independent of the environment surrounding the organism. The external physical environment, that is, the sum total of all the physical factors surrounding an organism, inevitably influences the internal biological environment of the organisms. Appropriate environmental conditions within the organisms (the physiological conditions) enabling them carry on their functions are struck by their physical environment. If the environmental factors change to a significant extent, the physiological changes occurring within the organisms cannot refrain from being altered.

Environmental physiology, thus, is the study of biochemical processes taking place in organisms and their coordination with the physical environment leading to various functions performed by the organisms. Functions performed by the organisms, in fact, are as a result of the processes going on within them. Environmental physiology, in other words, is the science of organisms' functions as influenced by their physical environment.

If environmental conditions change, physiological alterations ought to take place to certain extent and, then, the organisms tend to adjust their "internal" environment in such a way that their physiological conditions do not change significantly, and their growth and reproduction performances continue to be normal. However, drastic change in the operating environmental factors could be detrimental to the physiological conditions of the organisms. Under such circumstances, organisms tend to adapt to their environment, or they would perish. Adaptation of the organisms to their surrounding environment happens at the level of their physiological conditions. How organisms function in response to operating environmental factors, how they adapt to changing environmental conditions, and how the means by which the adaptation mechanisms evolved among living organisms are all studied under a distinctive biological discipline knows as environmental physiology. Alternately, also called ecophysiology or physiological ecology, environmental physiology is of crucial importance in our time because climate changes are posing a challenge to organisms within their changing ecosystems.

CLIMATE AND ITS CHANGING BEHAVIOR

Climate is the key player in determining the thermodynamics of the biosphere. Therefore, climate is the major force to influence evolutionary patterns of life. The patterns of plants' and animals' distributions are attributable to climate changes during the ice ages of the Pleistocene (the past 2–3 million years). Species populations have advanced and retreated as the climate changes have taken place. An analysis of the biological remains, particularly the buried pollens, led us to understand the vital influences of climate on the evolutionary patterns of life.

During the 20,000 years since the peak of the last glaciations, global temperatures have recorded a rise of about 8°C (Townsend et al. 2008). The biosphere has been "customarily" experiencing climate change since its evolution and this, indeed, has been a "game changer" of natural evolution. But climate change has been continuously experienced since the last quarter of the twentieth century, which is becoming dynamically more and more conspicuous now, appearing to be more serious in its nature than the previous ones. Greenhouse gases are increasing rampantly due to unprecedented anthropogenic activities. Carbon dioxide, the dominant greenhouse gas, alone has increased to the extent of 415 ppm.

Global warming seems to be setting unprecedented trends. An increase of about 8°C in the global temperature occurred over approximately 20,000 years during the postglacial period. But even this "mild" trend in temperature rise over a vast period could not prevent changes in the vegetation. The current trends in global warming reveal a record increase of 0.8°C in less than a half-century period. In some vulnerable ecosystems, such as the Himalayas, temperatures have risen more than 1°C. The consequences of climate change in the twenty-first century are predicted to be much more serious: an increase of more than 4°C in global temperature in 100 years. The predicted global warming is nearly 100 times faster than the postglacial warming (Townsend et al. 2008), and rapid climate changes in the future could result in extinctions of many species (Davis and Shaw 2001), and even the mass extinction of species could be expected if such a trend of global-warming-led climate changes goes on unabated.

A change in the climate phenomenally influences the physiology of organisms. Organisms first tend to adapt to climate changes. If the climate change is too drastic to allow organisms to adapt, they tend to migrate to the regions of favorable climate. Seasonal migration of several wild animals and birds and structural changes in natural ecosystems in tune with changing seasons are testimony of the phenomenal impact of the climate. Climate "behavior" results in characteristic growth patterns, reproduction, survival, and sustainability of living organisms and ecosystems. When climate behavior undergoes changes, normal patterns of life tend to undergo changes.

SCOPE OF ENVIRONMENTAL PLANT PHYSIOLOGY

Environmental plant physiology is a meaningful fusion of environmental science and plant physiology. A deteriorating environment all set to steadily culminate into life-threatening climate crises is the foremost fundamental issue facing our contemporary world. Resolving this issue is, undoubtedly, the greatest challenge facing humanity in contemporary times. Environmental physiology helps us deeply understand the complex environment–plant relationships at the physiological level. In other words, environmental plant physiology provides ample scope of understanding the phenomena of plant life vis-à-vis environment.

Environmental science is complementary to physiology. An understanding of environmental science is, in a way, a prerequisite for an understanding of environmental plant physiology. While environmental science examines environmental influences on plant physiology, the latter happens to be instrumental in examining the peculiar behavior of environmental factors as well as the state of the environment.

Energy, nutrient, water, temperature, altitudinal, and allelo-chemical relations of plants, physiological effects of stress (caused by water scarcity, atmospheric pollution, etc.), and climate change are major premises of environmental physiology. Environment–plant relationships are reciprocal, influencing the original composition and health of the biosphere (Gardner 2018). Knowledge of plant physiology can be utilized in industry in a big way (International Association of Plant Physiology 2003). Crises of the environment create a twist in these vital relationships. And a deviation in these relationships precipitates into socioeconomic and cultural crises for the humanity.

Some crucial basic life phenomena that environmental plant physiology helps us deeply understand are:

- Response of plants to various environmental factors, for example, light, temperature, winds, altitude, pollutants, and climate
- Energy utilization by plants, photosynthesis, chemosynthesis, energy budgets, etc.
- Nutrient requirements of plants, absorption, and transport of nutrients
- Water needs of the plants
- Seed germination, growth, flowerings, reproductive cycles and growth movements in plants
- Synthesis and breakdown of various metabolites in plant cells
- Plants' abilities to adapt to adverse environments
- Use of plant physiology as a tool in the mitigation of environmental pollution and climate change

APPLICATIONS IN FOOD PRODUCTION

Environmental plant physiology serves as an applied science in agriculture. It is a fast-emerging science applicable in influencing plant growth and functions vital for ameliorating agriculture. Knowledge of the environment-linked physiology is especially applicable in the fields of agronomy,

genetics, plant breeding, plant pathology, horticulture, forestry, and soil science. Some of the vital applications of environmental plant physiology in agriculture are:

- Knowledge of environmental plant physiology is important for choosing seasons and timings of crop plantation and making effective water management for agriculture.
- Physiological studies become the basis to determine doses of chemical and mined fertilizers and growth regulators for better growth of the crops.
- Soil-fertility management based on the applications of optimum amounts of fertilizers and root-microbe interactions can be ensured, applying the knowledge of plant physiology.
- Crop rotations, crop combinations, complementarity, and synergies between different crops/crop cultivars can be designed and exploited in accordance with various plant physiological parameters.
- Weed control and pest-management tips for plant protection and improved growth and productivity are readily provided by plant physiology.
- Tissue culture of some rare plants and micropropagation, which could solve numerous food-related problems and conservation of genetic resources, are the domains of plant physiology.
- Environmental plant physiological principles are efficiently applicable in greenhouse food production.
- Environmental plant physiology facilitates the selection of stress-, drought-, and disease-tolerant varieties.
- Production of seedless fruits, off-season fruits and vegetables, fruits and vegetables of long shelf life, flowers of long vase life, etc. have been possible thanks to the knowledge of plant physiology.
- Crops can be grown even in deserts by applying the principles of crop physiology, especially by monitoring water potential.
- Knowledge of environmental plant physiology provides ways and means for crop management under biotic and abiotic stress.
- Selection of seed varieties compatible with environmental/climatic conditions and in accordance with edaphic factors is made on the basis of the knowledge that plant physiology provides.
- Environmental plant physiology is pivotal in evolving and formulating adaptation strategies for agricultural production as per different agroclimatic zones, ecological zones, and in tune with the changing climate.
- Environmental plant physiology, in essence, is of key value for implementing sustainable agriculture and ensuring food security for humanity.

ENVIRONMENTAL PHYSIOLOGY: THE BASIS FOR A CLIMATE-SMART PLANET

Environmental plant physiology opens the floodgates to a treasure of knowledge applicable in resolving numerous issues pertaining to life as mentioned in the previous section. It also holds the key to arresting and reversing global warming and putting the climate back into normal order. Our biosphere is facing the grimmest crises ever: the climate change. Climate change is largely attributable to ongoing environmental disruptions. Environmental pollution, especially the rapid rate of carbon emissions into the atmosphere due to anthropogenic activities, is the major cause of climate change.

There are several issues posing a threat to life on the planet. Let us look at the burning questions environmental plant physiology promises to respond to:

- What is the nature of the relationships between the physical environment and the organisms?
- What happens when the operating environmental factors are extremely deficient or are in extreme excess?

- How are the physiological parameters affected by the operating environmental factors?
- When and how are the environment–plant relationships affected?
- What could be repercussions of extreme environmental conditions?
- How are various sorts of stresses developed in plants?
- What is the overall performance of plants (including food crops) attributable to?
- What could be the possible implications of a changing climate on the physiology and plant-production performance?
- How can the plants overcome environmental stress?
- What are the adaptation mechanisms helping the plants (including food crops) adapt to unfavorable environmental conditions, inhospitable environments, and changing climate?
- How can the climate-change mitigation processes be implemented and exacerbated?
- How can environmental plant physiology be instrumental in evolving strategies for a climate-smart and sustainable world?

Environmental plant physiology offers solutions to all the above-mentioned questions. And in the response to these questions is hidden much of the solution to climate crises. The first and the foremost phenomenon to combat global warming is the phenomenon that primarily utilizes atmospheric carbon dioxide and controls its atmospheric concentrations. It is photosynthesis: the phenomenon that brings the Sun down to Earth to flow through living channels. In photosynthesis, when passing through the "gateway" of the chloroplast, the sunlight synthesizes itself into life by fixing atmospheric carbon into plant biomass, and a carbon balance is struck. Photosynthesis sequesters atmospheric CO_2 into plant biomass and eventually into the soil. Soil organic matter and organic carbon build up due to photosynthesis and enriches soil ecosystems with life (largely the micro life), which, in turn, ameliorates soil fertility—a process of healthy pedogenesis. Fertile soil inhabits a variety of microorganisms, many of which play a crucial role in biological-nitrogen fixation, a process that further adds to soil fertility. More fertile soil consequently supports more and more photosynthesizers (green plants and photosynthetic microorganisms), which, in turn, further accentuate carbon sequestration and fortify the process of climate mitigation.

Chemosynthesis, pedogenesis, and photosynthesis together create a phenomenal "substance" comprising the rhizosphere, pedosphere, and phytosphere, respectively, integrated with each other and dependent on each other (Figure 1.11). There is continuous flow of air, water, and nutrients

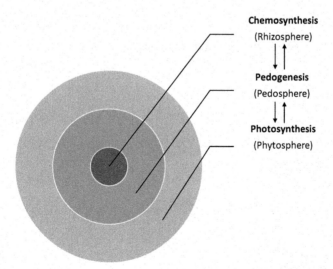

FIGURE 1.11   Chemosynthesis-pedogenesis-photosynthesis: A phenomenal "axis" of a climate-smart planet.

among these three components interrelated through three different phenomena. These three spheres fed by all the life-supporting factors available in the biosphere considerably influence plant physiology and also constitute a phenomenal basis for a climate-smart planet.

More of the life in the fertile soil habitats includes the chemosynthetic microorganisms. Chemosynthesis is a must for the plants to be nourished by nitrates as the basic source of nitrogen for the plants to enable them to synthesize their proteins. Proteins, as we know, are the structural constituents as well as functional biomolecules of life. Providing the only source of nitrogen to photosynthesizers, chemosynthesis itself nourishes the photosynthesis. Providing the basic source of nutrients (the organic matter), the photosynthesis itself nourishes the chemosynthesis. Chemosynthesis and photosynthesis, thus, are complementary to each other. Together these two phenomena create synergy for life to perpetuate throughout planet Earth in the biosphere.

In environmental plant physiology, we begin with the study of effects of light on living organisms, that is, the study of energy relations of life. The energy relations involve photosynthesis, a natural subject of environmental physiology. Energy relations also involve energy accrued from inorganic molecules and organisms' dependence on this energy, that is, chemosynthesis. Thus, this subject might be instrumental in dealing with the issues relating to climate change (Figure 1.12).

The biosphere is an energy system in which cosmic energy (largely the solar energy) transforms into "living" energy through photosynthesis. The more efficient and intensive the phenomenon of photosynthesis the more efficiently functional the energy system: the biosphere. An energy system also often attains entropy owing to disturbances in energy balances. In the biosphere this entropy is owing to disturbances between environmental (free) energy and "living" energy. Photosynthesis strikes a balance between the energy of the abiotic and biotic components of the biosphere and thus plays crucial role in combating the entropy.

Climate change is also due to disturbance in the energy system thanks to changes in the atmospheric composition. Emission of greenhouse gases in excessive amounts is culminating into global warming, contributing to altering the energy system leading to climate change.

Environmental plant physiology, unlike many other disciplines, absorbs an interdisciplinary agenda. It has entered the field of molecular biology and has interfaced with environmental science, ecology, environmental engineering, biophysics, biochemistry, microbiology, genetics and plant breeding, agronomy, forestry, horticulture, soil science, etc. With intrinsic relationships with so many disciplines, environmental plant physiology has evolved necessary tools and methodologies for investigation and development processes to promote growth, change genetic makeup, modify ecosystem structures, increase photosynthetic efficiencies, increase primary productivity,

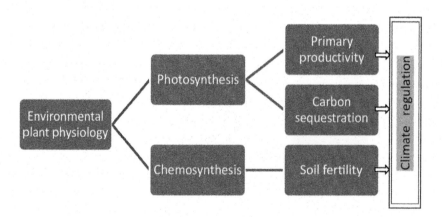

FIGURE 1.12 Climate regulation phenomenon: an attribute of environmental plant physiology.

enhance ecological regeneration, fortify ecosystem functions, and enhance carbon sequestration in the earth's ecosystems and offers great promises to strike environmental balances and ameliorate climate-mitigation processes indispensable for restoring a climate-smart planet.

SUMMARY

The universe is a lumenosphere, the home of light. The biosphere, a living oasis of the lumenosphere, is a phenomenon of light. Light is the only non-thing (energy) with power to synthesize life within the biosphere—through a process known as photosynthesis. The other phenomenon of life, known as chemosynthesis, involves the use of chemical energy in some inorganic molecules. Environmental physiology helps us understand the phenomenon of life in the biosphere.

A study of environmental physiology begins with the basics of ecology, a study of the reciprocal relationships of the living organisms with their physical environment and among themselves. This introductory chapter sets a background for discussion of various crucial aspects of environmental plant physiology. Ecological analysis pertaining to an ecosystem, an independent structural and functional unit of nature characterizing a community of interdependent organisms—plants, animals, microorganisms—and the ecological processes sustaining the life. Producers (photosynthesizers), consumers (herbivores and carnivores), and decomposers (fungi and bacteria) comprise the biotic components of an ecosystem. The abiotic components include the organic matter to emanate from dead plant and animal parts and inorganic components (water, nutrients, and gases). All ecosystems (barring chemosynthetic communities around the hydrothermal vents deep in the oceans and seas) are the solar-powered systems. Carbon dioxide, nitrogen, mineral nutrients, and water are the major inputs, and water, oxygen, and heat generated through respiration are the major outputs of an ecosystem. Water, in fact, is both, an input and an output, in the functioning of an ecosystem.

Energy (unidirectional) and nutrient (cyclic) flows within an ecosystem can be depicted by means of food chains and food webs. There are two types of food chains in an ecosystem, viz. grazing food chains and detritus food chains. A food web is a complex network of several food chains prevailing in an ecosystem. Producers and consumers make up five trophic levels in an ecosystem: producers, herbivores, primary carnivores, secondary carnivores, and top carnivores. Deriving energy and nutrients from all the five trophic levels, the decomposers constitute the sixth trophic level. Indiscriminate human intervention in natural ecosystems leading to a shift in the type of ecosystems—from natural to anthropogenic—suggests that human species serving as a major consumer of the ecosystem resources at every trophic level must be categorized as the seventh trophic level. According to the Second Law of Thermodynamics, there is gradual decrease in energy content along successive trophic levels of a food chain. At each transfer of energy through food, only 10%–20% is available to each successive trophic level. About 80%–90% of energy available at the lower trophic level is used in overcoming its entropy and in maintaining metabolic activities. Ecological pyramids are the graphical representations of the number, biomass, and energy parameters of different trophic levels. While the energy pyramid is always vertical or straight with producers at the base and top carnivores at the tip of the pyramid, the number and the biomass pyramids can be straight or inverted. The pyramid of numbers can also be of the spindle shape.

Environmental physiology is the study of biochemical processes taking place in organisms and their coordination with the physical environment leading to various functions performed by the organisms. Environmental physiology, in other words, is the science of organisms' functions as influenced by their physical environment and is of crucial importance in our times when the climate changes are posing a threat to the living organisms. A change in the climate phenomenally influences physiology of the organisms. Changes in the climate behavior lead to observable and often negative alterations in normal patterns of life. As a meaningful fusion of environment and physiology, the applied knowledge of environmental physiology has ample scope in our world, spelling out its attributes in food production, industry, and in almost every walk of life. Environmental plant physiology also serves as a key to arrest and reverse global warming and put the climate back into its normal order.

REFERENCES

Davis, M. B. and Shaw, R. G. 2001. Range shifts and adaptive responses to quarternary climate change. *Science* 292: 673–679.

Gardner, F. P., Pearce, R. B. and Mitchell, R. L. 2018. *Physiology of Crop Plants*. Jodhpur, India: Scientific Publishers. 327 p.

International Association for Plant Physiology. 2003. *Souvenir: Second International Congress of Plant Physiology on Sustainable Plant Productivity under Changing Environment*. New Delhi, India: ICAR. 152 p.

Lindeman, R. L. 1942. The trophic-dynamic aspect of ecology. *Ecology* 23(4): 399–417.

Singh, V. 2019. *Fertilizing the Universe: A New Chapter of Unfolding Evolution*. London, UK: Cambridge Scholars Publishing. 286 p.

Townsend, C. R., Begon, M. and Harper, J. L. 2008. *Essentials of Ecology*, 3rd edn, p. 59. Malden, MA: Blackwell Publishing. 510 p.

WEBSITES

http://www.biodiveridad.gob.mx/v_ingles/planet/whatis_bios.html.

https://blogs.timesofindia.indiatimes.com/toi-edit-page/photosynthesis-the-core-of-spiritual-karma/ (Singh, V. 2018. Photosynthesis: core of spiritual karma. *The Times of India*, June 18, 2018).

https://www.britannica.com/science/biosphere (Thompson, M.B., Thompson, J.N. and Gates, D.M., accessed on July 21, 2018).

https://www.speakingtree.in/article/let-the-sun-come-down-to-earth (Singh, V. 2017. Let the sun come down to earth. *The Speaking Tree*, June 18, 2017).

2 Energy Relations

The most part of leaves pour out the greatest quantity of this dephlogisticated air (oxygen) from their under surface, primarily those of lofty trees.

Jan Ingenhousz (1730–1799)

ENERGY AND LIFE

Living organisms are made up of matter and energy. The matter in organisms exists in all three states: solid, liquid, and gas. The solids comprise all the necessary elements called nutrients. The water is the major component of matter in the liquid state. Nitrogen, carbon dioxide, and oxygen are the three main gases all living organisms require in their structural composition. Carbon in living organisms gets fixed through photosynthesis (in chlorophyllous plants, algae, and cyanobacteria), chemosynthesis (in microorganisms), and organic molecules derived from organic foods (in fungi and animals).

The matter imparts structure (shape, size) to the organisms. Energy makes the matter function. As we know, it is not the matter that works; it is the energy that makes the matter work. Matter, in essence, is just a framework of an organism for the energy to perform specific functions in consonance with the matter. Matter and energy in living systems do not exist in isolation but are integrated into a conscious oneness. Matter cannot function without energy making it function. Energy dwells in the bonds of the biomolecules within the structures of the individual organisms. The matter-energy cartel imparts dynamics to organisms and, therefore, to life. The matter in an organism is never in a fixed amount. Due to metabolism (anabolism + catabolism), an organism exists in the states of wear and tear. In order to maintain this wear and tear, an organism regularly depends on the intake of all those molecules, which include all the necessary materials in alliance with energy. These molecules are combined in what is called as food. Thus, all organisms, and therefore the whole life, blossom due to food. Food is the source of nutrients and energy, the very sources of the existence, dynamism, and sustainability of life.

The energy source all living organisms on Earth use is in three forms, viz. light, inorganic molecules, and organic molecules. The organisms using sunlight directly are categorized as photosynthetic (photoautotrophs), and those using inorganic molecules are chemosynthetic (chemoautotrophs), and both together are referred to as autotrophs (Greek: *autos* = self, *trophe* = nourishing). The organisms that use organic molecules to derive energy for themselves are categorized as heterotrophs (Greek: *heteros* = the other, another, different, *trophe* = nourishing). Photosynthetic autotrophs (or photoautotrophs) directly use sunlight as a source of energy and CO_2 as a source of carbon.

TROPHIC DIVERSITY IN NATURE

We can't help appreciating the trophic diversity prevailing among the prokaryotes. The Prokaryote Kingdom of Life draws on all three sources of energy: light, inorganic molecules, and organic molecules. The unicellular organisms with no membrane-bound nucleus and organelles, which include bacteria and archaea (distinguishable from bacteria on the basis of their structures, biological features, and physiology, mostly occurring in the marine environments), are undoubtedly the most adapted organisms in the biosphere capable of harnessing all kinds of energy sources (Table 2.1). Existing, proliferating, and sustaining as photoautotrophs, chemoautotrophs, and heterotrophs, prokaryotes occupy all of the ecosystems in the biosphere, feeding on light directly, as well as on the

TABLE 2.1

Trophic Diversity Across All Living Beings on Earth

Living Kingdoms	Autotrophs		Heterotrophs
	Photosynthetic	Chemosynthetic	
Prokaryotes (Bacteria, Archaea)	+	+	+
Protists	+	−	+
Plants	+	−	+
Fungi	−	−	+
Animals	−	−	+

dead and living organisms and inorganic compounds, and "enjoying" the entire diversity of the nature. They are evolutionarily capable of fixing carbon using solar radiation (through photosynthesis) and energy in the inorganic molecules (through chemosynthesis), live in symbiosis with the organisms from other kingdoms, draw energy and nutrients from the living organisms of higher orders, and decompose the organic matter into all of its constituents. These roles of the prokaryotes place them in these categories: producers (photosynthesis and chemosynthesis), symbiotic (e.g., nitrogen fixation), parasites (causing a variety of diseases), and decomposers (causing mineralization). In the process of decomposition, the prokaryotes derive nutrients and energy from the dead organic matter.

Prokaryotic trophic diversity in marine ecosystems has revealed some interesting aspects of life. Béjà et al. (2000) discovered a different kind of energy production from light, which was bacterial rhodopsin, a light-absorbing pigment also found in the eyes of the animals, as also in bacteria and archaea. In addition to several functions, the rhodopsin in bacteria and archaea is also involved in the proton pump during ATP synthesis, that is, in energy production. The study further revealed that the bacterial rhodopsin is widespread through the oceans and that the light sensitivity is adapted to local variations in light quality (Béjà et al. 2001).

Protists—the organisms belonging to Kingdom Protista—are either photosynthetic or hetero-trophs. The protists are mostly unicellular eukaryotic organisms, but some like algae and seaweed are multicellular. The photosynthetic protists are so vital for aquatic life, for they provide them with shelter and oxygen.

The plant kingdom is well-known for fueling life with photosynthesis. However, there are some plants that defy photosynthesis because they are non-chlorophyllous and exist in the heterotrophic mode, deriving all or part of their nutrition from other organisms. Many parasitic, saprophytic, insectivorous, or carnivorous plants are examples of heterotrophic plants.

All organisms belonging to Kingdom Fungi, such as mushrooms, molds, and yeast, are hetero-trophs mostly dependent on dead organic matter for their nutrition by absorption.

The whole of Kingdom Animalia embracing an amazing diversity of animals—eukaryotic and multicellular without exception—are invariably heterotrophic, gaining nutrition from all trophic levels in the earth's ecosystems ranging from producers (plants) to all levels of consumers.

THE LIGHT OF LIFE

The living energy or the biochemical energy all living organisms work with, in essence, is the modified form of cosmic light—the light emanating from the stars (on Earth from the sun, the star nearest to the planet). Light enters into life by means of photosynthesis. All forms of energy on Earth, in fact, have their origin in light. Be it fossil-fuel energy or the energy in the bonds of inorganic and organic compounds, all energy forms are the transformed forms of sunlight. Chemosynthesis that uses chemical energy of inorganic compounds, in fact, is a slightly modified photosynthesis using alternative, rather than the direct, sunlight. The energy that keeps the animal kingdom going on is also the solar energy that was first fixed by the plants. The energy that operates metabolism of all organisms, and thus sustains the organisms, that is, adenosine triphosphate (ATP), is also a modified form of sunlight. We can call ATP the living light. Thus, the whole of the biosphere reverberates with living light. Our universe is a lumenosphere (or photosphere)—the Home of Light—and so is our biosphere. The only difference is this that in the biosphere the light keeps on transforming itself into life. Photosynthesis is the phenomenon of light with which light streams itself into life. Before we peep deeper into this wonderful phenomenon that is at the heart of the Living Planet, let us understand the nature of light.

Light travels through space as a wave with its specific frequency and wavelength. But when light interacts with matter, it acts as particles rather than as a wave. A particle of light is called as a photon. Each photon bears a finite quantity of energy. Infrared light, ultraviolet light, and visible light denote various wavelengths of light that matter to life. Infrared light is of longer wavelength and carries less energy compared to ultraviolet and visible light, which bear short wavelengths. The infrared light works to increase motion of the molecules leading to increased temperature of the matter. This light, however, does not contribute to drive photosynthesis because it does not bear as much energy as will be required for operating the mechanism. It merely contributes to maintain temperature regulation among living organisms. On the other hand, ultraviolet light is so rich in its energy content that it destroys the mechanism of photosynthesis by breaking down the covalent bonds of organic molecules.

In between the two extremes of the light spectrum is the light we see, and the light we see is indifferent with the light that drives photosynthesis. This is the visible light. The visible light and the photosynthetically active radiation (PAR) are the same. On solar spectrum, PAR falls between wavelengths 400 and 700 nm. This range of the light has an appropriate amount of energy capable of driving photosynthesis and preventing damage to organic molecules. To put it in a different way, PAR spells out its benevolence to life!

Of the total energy content of the solar spectrum at sea level, PAR makes up less than half (45%), and infrared at 53%, with ultraviolet light being the rest of the energy proportion (Molles 2005). Of the total radiation emitted by a terrestrial environment at typical terrestrial temperature, as much as 99% is the thermal radiation that lies in a region (above 4,000 nm) outside PAR (Nobel 1991; Christopher and Casper 2002).

What is fascinating to know is this that we (as well as the rest of the organisms with eyes to see) can only see that light that is fixed by photosynthesis. It is indeed photosynthesis that empowers us to see the light and everything in the light, with the help of the light that is so stupendously inoculated in us by the wonderful phenomenon of the light—photosynthesis.

Light falling on or entering into an ecosystem is not uniform in its quantity and quality. It changes with the time of a day, weather, and seasons at a place or in an area. It also changes from one region to another, in accordance with latitude, altitude, and slope orientation. Water bodies significantly bring changes in the quantity and quality of light. Even the organisms themselves are capable of inducing changes in light.

In water bodies, it is only the superficial euphotic zone that receives as much light as would be able to support photosynthetic organisms. The canopy of a dense natural forest absorbs about 79% of the total PAR falling on it, while plants in the middle layer absorb only some 7% of PAR. As light filters from the canopy of the forest, it is reduced to the extent of 1%–2% on the floor vegetation (Molles 2005). Transmission of light through a forest ecosystem, apart from geographical factors, will vary according to leaf types, leaf moisture content, leaf "waxiness," forest/tree developmental stage, etc. A forest also changes quality of light. Within the PAR range of the solar spectrum, the leaves absorb mainly blue and red light and transmit green light.

LEAF AND LIGHT

The leaf of a plant is not something ordinary. This is an extraordinary—rather the most extraordinary—structure worth being marveled. It is this structure on Earth that has the power to bring the sun down to Earth. A green leaf embraces the whole "machinery" that makes a bridge between the sun and Earth, between the inorganic universe and an organic biosphere. The leaf encompasses the biological power that transforms matter into spirit. The leaf is a miracle of the universe that transforms the non-living into living. The leaf is the original source of all food for all organisms directly or indirectly and is the source to nourish, evolve, and sustain the biosphere. The leaf is home to the most abundant enzymes on Earth, and the carbon cycle of nature pivots on it. The leaf is vital for the regulation of energy and temperature, and it operates the phenomenon of photosynthesis that feeds the whole biosphere directly or indirectly. The leaf serves as a vehicle for the movement of light from the sun to life.

A leaf, being a site of photosynthesis, is a key structure shaping the biosphere. A leaf, speaking philosophically, serves as a gateway for light to life. There exist as many as 391,000 species of vascular plants on Earth, comprising about 94% of the species belonging to the flowering plants. Every plant species has distinctive features of its leaves, the first and foremost feature imparting the morphological distinction to a plant species. So, we can imagine such an amazing diversity of "gateways" of light to life on Earth. This is such a fascinating aspect of life! The sun comes down to Earth to synthesize life in numerous ways via innumerable varieties of leaves following the same phenomenon—photosynthesis!

Morphology and anatomy of the leaves (Figure 2.1) are well adapted to what they perform: photosynthesis. The large surface area of a leaf is to absorb more light. A thin blade is good to facilitate quick absorption of carbon dioxide. A network of veins supports the leaf and transports water and carbohydrates. Stomata are for carbon-dioxide absorption and transpiration. A thin and transparent epidermis facilitates higher amounts of light reaching the palisade layer. A thin wax cuticle protects the leaf while not preventing light from being absorbed. The palisade layer at the top of the leaf helps the absorption of more light. The spongy layer increases the surface area and facilitates quick diffusion of carbon dioxide. Chloroplasts absorb the available light and transfer solar energy to chemical energy.

Leaf-Energy Balance

All living organisms, like all objects in nature, stay in interaction with their physical environments by means of an energy exchange. Metabolism in organisms operates at certain ranges of temperatures. For example, enzymes involved in photosynthesis function at some optimal temperature. If the temperature deviates beyond a limit, the photosynthetic enzymes will become denatured, blocking the function of the leaf. Therefore, organisms attempt to maintain equilibrium either by gaining or losing energy.

The leaf of a plant is a logical starting point to understand how an organism, ecosystem, or the entire biosphere stays in equilibrium. When a leaf is in equilibrium, energy input equals energy output. If the leaf is not in equilibrium with its environment, it means the temperature of the leaf will change—increase or decrease—until equilibrium is attained. It takes less than a minute for a leaf to attain equilibrium. There are certain points to ponder in order to marvel at the fascinating role of a leaf in striking an energy balance:

- Most of the sunlight is intercepted by leaves before it reaches the ground.
- In the dense forests and grasslands, only about 2% radiation reaches the forest floor.
- As much as 95% of the sunlight absorbed by leaves is converted into heat, ending up in increasing leaf temperature. Even if the environmental conditions are ideally appropriate, only a very small fraction of the absorbed light is converted into chemical energy through photosynthesis.
- Evaporation of water takes place from the cell surface by means of transpiration, a process that requires energy from the leaf, thereby reducing leaf temperature.
- Water stress in plants occurs thanks to limited availability of water in the soil, coupled with excessive evaporative demand in the environment.

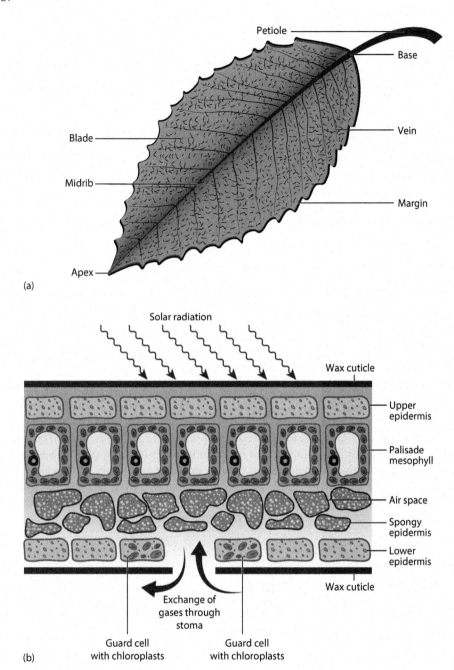

FIGURE 2.1   (a) External (above), and (b) internal structure of a typical leaf.

- In the event of water stress, stomata are closed, CO_2 intake is cut off, and photosynthesis slows down at a faster pace.
- Sunlight—that is, the energy source of photosynthesis—does photosynthesis, allowing fixation of CO_2, but also imposes the burden of water loss due to imposed heat load.
- How fast the photosynthesis occurs and how quickly the heat is lost to the environment is affected by the size, shape, and chemical contents of the leaves.

Energy Inputs and Energy Outputs of a Leaf

A leaf is exposed to a number of environmental factors. The interaction of the leaves with their environment encounters an energy input in the form of short-wave and long-wave radiation, which results in the creation of many energy outputs that are involved in striking an energy balance enacted by the leaves. The following are the inputs and outputs, the factors that matter in maintaining energy balance by leaves (Figure 2.2):

Inputs
* Short-wave and long-wave radiation

Outputs
* Long-wave radiation
* Conduction and convection
* Transpiration
* Transmission
* Reflectance
* Energy in carbon compounds (sugars)

These can be grouped into:
* Net radiation
* Exchange of sensible energy (heat)
* Latent energy (conversion of water into water vapors)
* Chemical energy (photosynthate)

Net Radiation

Energy absorption characteristics demonstrated by all leaves are generally alike. The long-wave radiation is an input as well as output. Most of the radiation is absorbed by plant leaves, but a fraction of it is reflected. Of the absorbed radiation, only that in a narrow set of wavelengths of the visible spectrum is converted into simple sugars by chlorophyll molecules, the first products of photosynthesis. The energy that is reradiated from the surface of the leaves is in long wave, that is, thermal, which increases with the fourth power of temperature in degrees Kelvin.

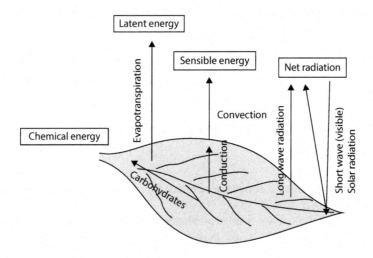

FIGURE 2.2   Energy balance of a radiated leaf.

Sensible Heat Loss (Conduction and Convection)

Conduction is the material's (leaf's) ability to conduct heat from one side to the other, as a function of the gradient in temperature. Convection is the loss of energy from a heated surface warming the surrounding air, and then removal by turbulent air flow. Some structural characteristics of the leaves, for example, size, shape, degree of lobing, or surface roughness, greatly affect the size of the "boundary layer."

- Thicker leaves have more mass to conduct heat to the edges, leading to a rapid loss of heat.
- Needle-shaped leaves of the coniferous plants generally tend to stay in the same temperature, similar to that of the surrounding air, because such leaves lose heat at a very rapid pace.
- Leaf hairs (often occurring in plants in dry areas) are efficient at the rapid conduction of heat and also increase reflectance and air turbulence, thereby contributing to a rapid loss of heat.
- Rapid loss of heat is also encountered in smaller, thicker, and more deeply lobbed leaves.
- The "dancing" peepal (*Ficus religiosa*) leaves are well-adapted to rapid heat loss.

Latent and Chemical Energy

A certain amount of water is lost through water vapor into atmosphere against CO_2 uptake by the plants. This occurs through the stomata in the leaves. This transpiration is a very crucial component of the leaf-energy balance. This requires latent heat from the leaf, resulting in a decline in leaf temperature.

Transpiration further requires movement of water from soil through the root and trunk to leaves of a plant in order to replace the water lost through water vapor.

Water stress, eventually followed by stomatal closure and significant reduction in water loss as well as CO_2 uptake for photosynthesis, is developed if water uptake by the plant is not enough. Thus "water cost–carbon gain" is an inevitable part that photosynthesis plays. The weight of carbon gained per unit weight of water lost to the atmosphere is what is called water-use efficiency (WUE):

$$WUE = \frac{C_{gain}\,(g)}{H_2O_{lost}\,(g)}$$

WUE can be defined at the level of a leaf or a plant, or at the level of a whole ecosystem. This is one of the key measures to understand photosynthetic efficiency as performance of the plants, especially of the food crops, dependents on irrigation.

Chemical energy in the budget of a leaf is affected in the following ways:

- A photosynthetic response curve summarizes the relationship between light intensity and rate of CO_2 fixation by a leaf. If intensity of light gradually increases over a leaf, the rate of photosynthesis gradually increases and then levels off.
- The maximum rate of photosynthesis is referred to as the "light saturated" rate.
- A higher rate of photosynthesis is observed in the species occurring under full-sun conditions than those thriving in partial shade.
- CO_2 fixation doesn't begin taking place immediately after a leaf is irradiated. Continuous respiration by cells releases more CO_2 than can be fixed during photosynthesis. It implies that CO_2 fixation is zero at some light level above zero. It is known as the "compensation point," which is higher for the plants with a maximum photosynthetic rate.
- High compensation points, as well as high maximum net photosynthetic rates, vary from species to species and are generally attributable to leaf morphology. Chlorophyll and nitrogen concentrations also affect these attributes.

- The plant species occurring in sunny conditions bear leaves with two layers of palisade cells bearing chlorophyll. There is also a higher density of chlorophyll per unit of leaf surface area in these species.
- Amounts of chlorophyll and that of proteins are positively correlated in the leaves, which seems to be essential for carrying out all the biochemical reactions of photosynthesis.
- Temperature plays a very crucial role in determining the rates of net photosynthesis because all the enzymes involved in the complex biochemistry of photosynthesis play their specific role under appropriate temperature ranges.

PHOTOSYNTHETIC PATHWAYS

Photosynthesis is a lively phenomenon in which the biosphere came into existence. This is the phenomenon that sustains life on Earth, making it a Living Planet. In photosynthesis, in essence, light transforms itself into life. The process, however, is not so simple. It involves a complex series of biochemical reactions in which CO_2 is converted into carbohydrates—a reaction that involves the reduction of carbon with H_2O as a reductant.

The chemical reaction called reduction means a gain of electrons, or gain of hydrogen, or loss of oxygen. The reverse of reduction is the chemical reaction called oxidation involving the removal of electrons from an atom, ion, or compound, or removal of hydrogen, or the gain of oxygen.

In photosynthesis, the reduction of a substance involves oxidation of the other. Photosynthesis is represented by one of the following reactions:

$$6CO_2 + 6H_2O + \text{Light energy} \rightarrow C_6H_{12}O_6 + 6O_2$$

$$CO_2 + 2H_2O + \text{Light energy} \rightarrow \left[CH_2O\right]n + H_2O + O_2$$

During photosynthesis, the photosynthetic pigments in plants, algae, or bacteria absorb solar radiation and transfer the energy to electrons to be used to synthesize ATP and NADPH, which, in turn, serve as donors of electrons and energy for the synthesis of sugars (Molles 2005). Two reactions, viz, the light and the dark reactions, make up photosynthesis, as outlined in Figure 2.3. Thus, the chlorophyll-containing organisms are capable of converting the electromagnetic energy of sunlight into energy-rich organic molecules that serve as a fuel to feed most of the biosphere and sustain it.

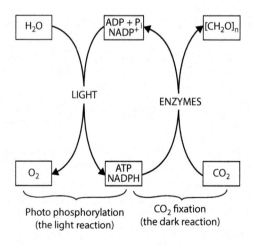

FIGURE 2.3   The light and dark reactions of photosynthesis.

Photosynthesis is often referred to as carbon fixation—the biochemical reactions incorporating CO_2 into a carbon-containing acid. All the photosynthetic organisms, however, do not carry out this energy conversion in the same manner. There are three different biochemical pathways in which photosynthesis is accomplished: C_3 photosynthesis, C_4 photosynthesis, and crassulacean acid metabolism (CAM) photosynthesis. These pathways are broadly determined according to ecological differences among photosynthetic organisms. The organisms following these three pathways are referred to as C_3 plants, C_4 plants, and CAM plants, respectively.

C_3 Photosynthesis

In C_3 plants, CO_2 is fixed following the Calvin cycle. CO_2 first combines with a five-carbon compound known as ribulose bis-phosphate (RuBP). Catalyzed by the enzyme RuBP carboxylase, the first product of the carbon-fixation reaction is a three-carbon acid, phosphoglyceric acid, or 3-PGA (Figure 2.4). It is because of the three-carbon molecules of PGA that it is called C_3 photosynthesis and the plants employing it the C_3 plants.

This role of CO_2 entry and exit are enacted by the leaf stomata. During the process of photosynthesis, as CO_2 enters, water vapor exits. However, the rate of CO_2 entry is slower than the exit of water vapor. Furthermore, RuBP carboxylase has a poor affinity for CO_2. It is because the water-concentration gradient between the leaf and atmosphere is steeper than that of CO_2 between the atmosphere and leaf. In other words, the CO_2 intake for photosynthesis costs some water loss. Such a loss is taken care of by the photosynthesizing plants, provided there is no scarcity of water and the climate is not dry and hot. High rates of water loss at the expense CO_2 intake can lead to stomata closure and photosynthesis shutdown.

What happens when stomata closure takes place under very hot and dry conditions to prevent excessive water loss? Under those circumstances, the CO_2 concentration in the chloroplast drops below about 50 ppm, and relative concentration of O_2 increases. Then the catalyst RuBisCO (ribulose-1,5-biphosphate carboxylase/oxygenase) begins to fix O_2 rather than CO_2. The reaction with CO_2 is the first step in the Calvin cycle, leading to sugar synthesis. However, when RuBisCO reacts with O_2, it is the first reaction of the photorespiration pathway "undoing" the Calvin cycle and expending energy (Figure 2.5).

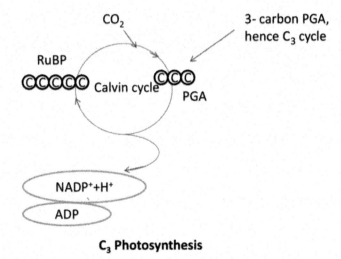

C_3 Photosynthesis

FIGURE 2.4 C_3 photosynthesis.

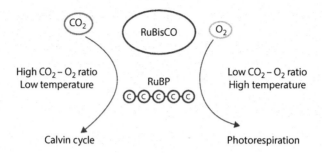

FIGURE 2.5 Dual action of RuBisCO.

The plant kingdom of the earth is dominated by C_3 plants. These include most of the vascular plants. All trees, all algae, and most of the food crops, including most significant cereal crops, legumes, and oilseed crops, belong to the category of C_3 plants. The bulk of the foods that humanity depends on is produced by C_3 photosynthesis.

C_4 Photosynthesis

C_4 photosynthesis is an alternative pathway in which carbon fixation and light-dependent reactions of photosynthesis take place in separate cells. In the C_4 plants, carbon fixation in combination with phosphoenolpyruvate (PEP) is accomplished in mesophyll chloroplasts, producing a four-carbon intermediate, typically malate (malic acid). This initial reaction is catalyzed by PEP carboxylase and contributes to concentrate CO_2.

PEP carboxylase has a very high affinity with CO_2 due to the internal concentration of CO_2 that is brought to very low levels. This situation leads to an increased CO_2 gradient from atmosphere to leaf, which, in turn, increases its inward flow and consequent consumption rates.

The acids produced in the mesophyll cells diffuse to specialized cells known as the bundle sheath, wherein the 4C acids are broken down to 3C acids and CO_2. A high concentration of CO_2, thus, is built up in the bundle-sheath cells, resulting in increased efficiency of the RuBP carboxylase enzyme that combines CO_2 with RuBP to produce PGA (Figure 2.6).

The C_4 plants remain nearly unsaturated with light and with limited water outperform C_3 plants under hot and relatively dry conditions. Out of some 250,000 terrestrial plant species, only 75,000 species—that is, about 3%—perform C_4 photosynthesis (Sage et al. 1999). Most of the

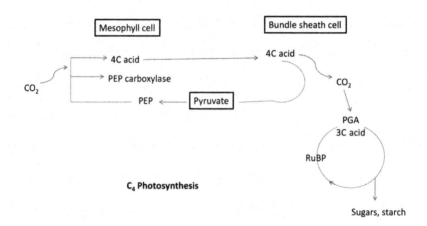

FIGURE 2.6 C_4 photosynthesis.

C_4 plants include grasses (4,500 species) followed by 1,500 species of sedges and 1,200 of dicots. Despite a limited proportion of all the plants participating in this type of photosynthesis, as much as approximately 25% of the primary productivity of the biosphere is attributable to this alternative pathway of C fixation (Lloyd and Farquhar 1994; Brown 1999). A large proportion of primary production for human consumption is derived from C_4 crops, which include food crops of critical significance, like sugarcane, corn, sorghum, and millets. Sugarcane, recorded at 7% photosynthetic efficiency, of course, is a champion of the C_4 photosynthesis. Large populations of livestock also subsist on pasture grasses performing C_4 photosynthesis. The C_4 plants, because of high water- and nutrient-use efficiencies, are capable of growing even in the habitats that are too harsh for C_3 plants. According to Sage (2004), it is because of C_4 plants that complex ecosystems existed, which might have otherwise been bare grounds.

CAM PHOTOSYNTHESIS

CAM is a sort of strategic photosynthesis adapted to extremely hot and dry environments. Representing some 10% of the total plant species, the CAM plants include cacti, some orchids and ferns, pineapple, and Spanish moss.

In the CAM photosynthetic pathway, carbon fixation takes place at night when, during CO_2 uptake, water loss at lower temperatures is reduced to a minimum. CO_2 taken in combines with PEP to form 4C acids, which get stored in the large vacuoles of their photosynthetic cells. With the 4C acid (malic acid) concentration up to 0.3 M, the pH drops to 4. During the daytime, the stomata are tightly closed to prevent water loss, and the 4C acids are decarboxylated to release CO_2, which then enters the Calvin cycle, that is, the C_3 photosynthetic pathway (Figure 2.7).

In CAM plants, PEP is used for the initial short-term C fixation, as also a case in C_4 plants. However, unlike in C_4 plants in which reactions take place in different cells, in the CAM plants, the entire chain of reactions goes on in the same cell. CAM strategy separates the reactions only in tune with time: initial C fixation at night, and 4C acids to Calvin cycle during daytime.

Although CAM plants do not normally perform photosynthesis at very high rates, their WUE (mass of CO_2 fixed per kg of water used) exceeds that of C_3 and C_4 plants.

Separation of the initial carbon fixation from other reactions helps reduce water losses during photosynthesis. Water losses for every gram (on dry-matter basis) of tissue produced during

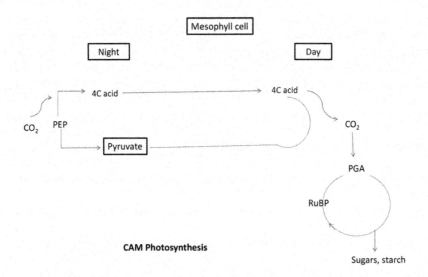

FIGURE 2.7 CAM photosynthesis.

C_3, C_4, and CAM photosynthesis are 380–900 g, 250–300 g, and 50 g, respectively (Molles 2005). Thus, among all the tree photosynthetic pathways, WUE is at a minimum in the case of C_3 plants, moderate for C_4, and maximum for CAM plants.

Be it C_3, C_4, or CAM photosynthesis, all the photosynthetic organisms (green plants, algae, and photosynthetic bacteria) capture carbon from CO_2 and energy from sunlight, packaging them in organic molecules. These are the molecules in the plants the consumer animals are dependent upon for deriving their carbon and energy as a source for their sustenance. Thus, the photosynthesizers opened gateways of evolution for the organisms (of consumer trophic levels) that are able to subsist and sustain on the energy and carbon of organic molecules.

$E = mc^2$ IN THE CONTEXT OF EARTH'S ECOSYSTEMS

The First Law of Thermodynamics states that energy can neither be created nor destroyed; it can only be converted from one form to the other. In Earth's ecosystems, solar energy is transformed into biochemical energy (through photosynthesis) and subsequently into heat energy (through respiration). Einstein's famous classical equation, $E = mc^2$, demonstrates that matter and energy are the same and that both can be converted into each other using the constant c^2, that is, the speed of light ($c = 2.99792458 \times 10^8$ m/s) multiplied by itself. This most-celebrated equation of the twentieth century (that led to the precipitation of mushroom clouds in Hiroshima and Nagasaki and operations of nuclear reactors as well) unfolds the matter–energy relationship on the surface of stars and in nuclear reactions anywhere in the cosmos and on Earth.

In Earth's ecosystems, the matter–energy relationship is of a different nature. Photosynthesis fixes solar energy into C-H bonds of organic molecules, such as carbohydrates, proteins, and fats. There are two basic and unique characteristics of Earth's ecosystems, viz.:

1. Matter is neither created nor destroyed; and
2. Energy conservation takes place.

The energy conservation in and by ecosystems (in species, populations, and communities) is basic and vital for the very existence and sustenance of life in the biosphere. The energy flows amongst various trophic levels, and thus, energy conservation creates enormous biodiversity in the biosphere. As the energy is fixed into biomolecules of the cells, it is used in synthesizing living matter within organisms' bodies and thus imparts size and shape to the organisms and through them to the ecosystems, and through ecosystems, to the whole planet. This energy also makes the organisms and all ecosystems function, allowing them to play their specific lively roles within the biosphere.

Primary productivity of the ecosystems is directly dependent on photosynthesis, while the secondary productivity is an outcome of nutrient-energy flows from producers to different trophic levels. In an ecosystem, there is a considerable exchange of matter between physical environments (soil–water–air) and within all living organisms. Changing from one form to another, matter in the biosphere keeps on recycling itself but remains constant in its amount, that is, neither created nor destroyed. Energy flow is unidirectional—from the sun to the photosynthetic organisms (green plants, algae, and cyanobacteria). In the biosphere, energy doesn't recycle. Once utilized in the synthesis of biomolecules and in operating and maintaining metabolism in organisms, when it changes into heat, it gets released into the physical environment and slowly exits to outer space (Figure 2.8).

The energy that is involved in photosynthesis and subsequent biomolecule synthesis in the biosphere is not being produced from the matter in the ecosystems (in its origin, that is, the sun, it is generated from matter as also in case of all the innumerable stars in the cosmos and, of course, in nuclear reactions managed by human beings and in the interior of planet Earth). The matter in the ecosystems is free from the "creation–destruction" cycle. The energy that performs its creative functions in Earth's ecosystems, beginning with photosynthesis, is an alien factor (making entry

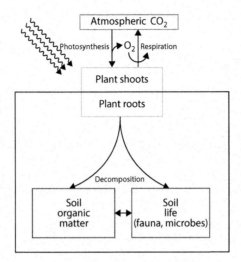

FIGURE 2.8   The matter–energy conservation cycle within an ecosystem.

from outer space), and hence the matter–energy relationship in the living systems is different from what it ought to be in outer space. Therefore, Einstein's most-sung equation $E = mc^2$ does not hold its validity in living systems.

The Ecological Law of Thermodynamics (ELT) and the thermodynamic concept of "exergy" introduced by Jorgensen (1992) underline the fact that "ecosystems are irreducible systems." Here, solar energy is "fixed" into "living energy" in the bonds of the earth's matter (CO_2 and H_2O). Upon the decomposition of organic matter, which Singh (2019) refers to as "dephotosynthesis," the energy that releases into an ecosystem, subsequently to exit to the space, is not yielded by the conversion of matter into energy. This is released from the matter it had synthesized via photosynthesis. Energy in ecosystems spells out its benevolence by being involving in the synthesis of life, the conservation of life, the physiology of organisms, the maintenance and sustenance of life, and in building up a climate system.

SUN–SOIL–PLANT: THE EVOLUTIONARY LADDER OF LIFE

The phenomenon of photosynthesis prevails on Earth, giving it a status of a Living Planet, the lone Living Planet we know about so far. Photosynthesis, however, is not just a terrestrial phenomenon. It is a universal phenomenon, for it involves light, the solar radiation, which is a non-terrestrial factor. Light—that is, the cosmic energy—seeds the earth. The phenomenon through which the light reveals its infinite benevolence is photosynthesis. Photosynthesis should not be understood only in terms of its chemistry but also in its universal implications: universal power of light to infuse life into non-living matter.

The universal sun–soil–plant connect mediated by photosynthesis is one of the most vital phenomenal outcomes of cosmic evolution—the phenomenon that imparts a dynamic dimension to the universe through wear and tear and the renewal of everything. Photosynthesis converts light energy into living (biochemical) energy, accumulating matter in the form of biomass that imparts innumerable shapes and sizes to living organisms and produces an oxidizing agent—the oxygen—that further breaks down biomolecules, making them perform various functions, including growth, motion, and reproduction.

Photosynthesis doesn't perform its functions linearly and in a simple manner. It performs in a very complex manner, as we have understood above, and creates for itself an extremely fertile ground vital for it to blossom into life in its richness. This ground is the soil. The soil is an "organic home" for innumerable living organisms—mostly microorganisms—and this fertile "home" further supports another set of innumerable living organisms (an amazing diversity of species), that is, the plants. Thus, photosynthesis constantly mediates the sun–soil–plant connect. This connect exists, and life is sustained on Earth. The higher the efficiency of photosynthesis the stronger the sun–soil–plant connect.

Photosynthesis and Soil Fertilization

Photosynthesis is the original source of soil fertilization. Organic fertilizers are the products of photosynthesis. Total organic carbon (TOC) stored in soil organic matter (SOM) is a product of photosynthesis. Decomposition of plant and animal residues, plant-root exudates, soil biota, and living and dead microorganisms provides the way to organic carbon to enter the soil. Non-decomposed plants and animal residues make up the organic fraction, SOM, which is a heterogeneous and dynamic substance varying in C content, particle size, decomposition rate, and turnover time. In SOM resides soil organic carbon (SOC), which serves as the main energy source for soil microorganisms.

SOM size and breakdown rates determine the kind of SOC. An active pool of SOM is easier to breakdown (Table 2.2). SOM contains approximately 58% C. Thus, a factor of 1.72 is used to convert SOC to SOM.

SOC happens to be one of the most crucial fractions of the soil, for it serves as a source of energy and nutrients. Continuous availability of nutrients for the plants is due to mineralization of SOC. SOC helps microorganisms to proliferate, a process that also helps boost nitrogen fixation in the soil. Humus fraction imparts aggregate stability to the soil and enhances its nutrient and water-holding capacity (Reicosky 2007).

Polysaccharides, such as sugars, found in organic carbon help bind mineral particles into micro-aggregates. SOM has a substance called glomalin that contributes to stabilizing the soil structure by gluing aggregates together and also making the soils resistant to erosion while maintaining porosity for movement of air, water, and roots through the soil. Organic acids, such as oxalic acid, released as a result of the decomposition of organic residues, phenomenally helps in preventing PO_4 fixation by clay minerals, thereby enhancing PO_4 availability to the plants.

When SOC is in a diminished amount, as often is the case in cultivated lands suffering from continuous deprivation of organic matter as an external input, energy supplies to soil life are cut off and, as a consequence, microbial populations, activities (functions)—for example, nutrient mineralization and nitrogen fixation—are severely affected. Scarcity of SOC also leads to the breakdown of detritus food chains, reducing diversity in the soil biota and deterioration of the soil environment. An accumulation of toxic substances and an increase in pests and diseases in food crops are also the natural alarming consequences of reduced SOC.

TABLE 2.2
Size and Breakdown Rates of the Soil Organic Matter Fractions

Fraction of the Soil Organic Matter (Specialty)	Particle Size (mm)	Approximate Turnover Time (Years)
Plant residues (parts of the plants that can be identified)	Less than or equal to 2.0	5
Particulate organic matter (partially decomposed plant parts)	0.06–2.0	100
Soil microbial biomass (bacteria and fungi)	Variable	3
Humus (ultimate stage of decomposition mainly with stable compounds)	Less than or equal to 0.0053	100–5,000

Except humus, all the SOM fractions constitute an active pool relatively easy to break down.

The carbon pool in the soil, without which soil is just an inanimate substratum, is built up by photosynthesis. Photosynthesis first sequesters atmospheric carbon into plant biomass, and from there it continuously keeps on moving to the soil. With this carbon wealth, the soil is imbibed with a wonderful richness of life. Soil is not just a substratum to help plants anchor their roots. It is an ecosystem—not just an ordinary ecosystem but, undoubtedly, the most stable, most vibrant, and most wonderful ecosystem on Earth. Life in the soil is invisible, but its impact on the visible terrestrial life is phenomenal. The richness of the visible above-soil terrestrial life is directly proportional to the richness of soil life.

The productivity of our croplands is directly proportional to the productivity of soil ecosystems. Therefore, a plant grows to its potential functions to spell photosynthesis and life, when it is sown in the soil and fertilized by itself via photosynthesis. In other words, our food security is rooted into healthy and vibrant soil, and our food sovereignty is based on the life-laden soil environments. Food security is the primary indicator of the socioeconomic and cultural progress of a nation. Therefore, in essence, human happiness is rooted into soil. Our sustainable future lies in the soil. Since the soil ecosystem, like the above-soil terrestrial ecosystem, itself is a function of photosynthesis, all human progress, hope, happiness, and sustainability of the future lie in photosynthesis. Photosynthesis, in essence, is the soul of the sun–soil–plant relationship that makes life flower on Earth.

Soil, in essence, is the basis of ecological affluence on Earth. Human progress is also rooted in fertile soil. There is a saying: If you own soil, you own up to the sky. The ethical onus on the human species that evolved as a custodian of whole life is: Let the photosynthesis go on with its potential efficiency, and we shall own the whole cosmos.

ENERGY FROM INORGANIC MOLECULES

The organisms—only among the microorganisms—that are capable of deriving energy from the reactions involving inorganic molecules to synthesize complex organic compounds are the second category of autotrophs: the chemoautotrophs. This phenomenon occurring in the absence of light is known as chemosynthesis. Fixing carbon using the oxidation of chemical compounds, the chemosynthesizers are phylogenetically diverse, also representing biogeochemically crucial taxa, for example, gammaproteobacteria, epsilon proteobacteria, aquificales, neutrophilic iron-oxidizing bacteria, and methanogenic archaea.

The term "chemosynthesis" was coined by Wilhelm Pfeffer in 1897. Before that, in the 1880s, Sergei Nikolaevich Vinogradskii brought to the fore the fact that some microbes were able to live solely on inorganic substances and in 1890 proposed a novel type of life process that he called "anorgoxydant." The "anorgoxydant" later on was known as chemosynthesis and the process of food production through chemosynthesis as chemoautotrophy. In the 1940s, André Lwoff used the term "chemotrophy." Nowadays, this is also known as "chemolithoautotrophy."

Some chemosynthesizers oxidize hydrogen (H_2), carbon monoxide (CO), iron (Fe^{2+}), ammonium (NH_4^+), and nitrite (NO_2^-). Almost all the biological nitrogen fixation occurs due to the role of certain chemosynthetic microorganisms inhabiting soils, waters, and roots of some plants. These nitrifying bacteria are of very high ecological importance. Much of the structure of our biosphere is attributable to the nitrogen-fixing microorganisms. These bacteria oxidize ammonium (NH_4^+) to nitrite (NO_2^-) and nitrite to nitrate (NO_3^-) and derive energy for their own food production.

$$2NH_4^+ + 3O_2 \rightarrow 2NO_2^- + 4H^+ + 2H_2O + Energy$$

The most glaring examples of ecosystems based on chemosynthesis are around the hydrothermal vents on ocean floors. Before the great discovery of these wonderful ecosystems in the Galápagos

Rift in the Pacific Ocean in 1977, it was conceived that it was photosynthesis that empowered all ocean life. Discovery of the chemosynthetic communities amidst the total absence of light deep in the Pacific Ocean baffled ecologists. The energy captured by chemosynthesis being the very basis of life on deep ocean floors further helped marine ecologists to examine the rich communities and their feeding modes in absolute dark zones of the oceans.

Hydrothermal vents on the ocean floors, as a result of volcanic activity, discharge warm water with sulfur compounds, which are used as sources of energy by chemosynthetic bacteria. Chemosynthesizers derive energy by oxidizing elemental sulfur, hydrogen sulphide, or thiosulphite oozing out of the hydrothermal vents. The energy from the sulfur compounds is used by the sulfur-oxidizing bacteria in fixing carbon from carbon dioxide dissolved in ocean water.

$$2H_2S + O_2 \rightarrow 2S^0 + 2H_2O + \text{Energy}$$

$$6CO_2 + 6H_2O + 3H_2S + \text{Energy} \rightarrow C_6H_{12}O_6 + 3H_2SO_4$$

The sulfur-oxidizing bacteria are of two types, viz., free-living and inhabiting the tissues of several invertebrates. Among these invertebrates are the oft-referred giant tube worms (*Riftia pachyptila*), which are up to 4 m in length and a little over 4.0 cm in diameter (Figure 2.9). They have no mouth, no eyes, no legs, and no digestive tract. These marine invertebrates belong to the family of polychaete annelid worms. Since the fascinating discovery in 1977, more than 300 species of these unusual tube worms have been identified. The animal has green-brown spongy tissue called trophosome, accommodating specialized cells packed with chemosynthetic microorganisms. More than 10 billion of the microorganisms reside per gram of trophosome. The sulfur-oxidizing bacteria contribute to as much as 60% of the worm's total biomass. The upper part of the tube worm's body is called a plume. The bright-red color of the plume is due

FIGURE 2.9   The giant tube worm (*Riftia pachyptila*) from the hydrothermal vents at the East Pacific Rise at 2,500 m depth. Each individual giant tube worm in the image is more than 1.0 m in length. (Courtesy of Monika Bright, University of Vienna, Austria).

to hemoglobin. The plume takes up H_2S and O_2 with the aid of hemoglobin and derives energy to utilize carbon from CO_2 to synthesize biomolecules. In this way, the giant tube worm prepares food for itself, harnessing the symbiotic relationship with the chemosynthesizers. The outer covering of these invertebrates is made up of chitin, allowing them to survive in the extreme heat of the thermal vents. The giant tube worm is adapted to withstand extreme pressure and temperature ranges (boiling to freezing).

The chemosynthetic communities dependent on sulfur-oxidizing bacteria also prosper around the thermal vents in deep freshwater lakes, in caves, as well as in surface hot springs.

ENERGY FROM ORGANIC MOLECULES

All heterotrophs sustain themselves, deriving nutrients and energy from organic molecules. These organic molecules originally are a "product" of photosynthesis (we shall not discuss here the organic molecules synthesized through chemosynthesis). Organic molecules in plants used as foods emanate from the primary productivity of an ecosystem, while those in animals constitute the secondary productivity of an ecosystem.

The first category of heterotrophs includes all the animals feeding directly on primary produce, that is, the products of photosynthesis. These are called primary consumers or herbivores. The second category includes all the carnivores that feed on herbivores, on primary carnivores, or on secondary carnivores, referred to as primary carnivores, secondary carnivores, and tertiary or top carnivores, respectively. The other category of organisms is those that feed on non-living organic matter, the detritus. These are called detritivores.

CHEMICAL COMPOSITION VIS-À-VIS NUTRIENT REQUIREMENTS

All the living organisms depend on food that must meet their structural requirements. In other words, the chemical composition of the organisms in reference, and that of the food, must be quite alike, that is, the nutrients available in the food must meet the chemical composition of the organisms. Therefore, first of all, we should have a look at the chemical composition of the organisms and that of the foods they depend on. There are just five constituents, viz., carbon, hydrogen, oxygen, nitrogen, and phosphorus—or CHONP—that make up 93%–97% of the biomass of all organisms on Earth (bacteria, protozoa, fungi, plants, animals). Among all the five living kingdoms, plants are chemically the most distinctive, often containing lower contents of nitrogen and phosphorus. For example, plants contain 2% nitrogen, on average, whereas bacteria, fungi, and animals have 5%–10% nitrogen, on average (Molles 2005). There are many other nutrients that would make up the rest of the proportion (3%–7%) of the organism's biomass; these are calcium (Ca), magnesium (Mg), sulfur (S), chlorine (Cl), iron (Fe), zinc (Zn), copper (Cu), manganese (Mn), boron (B), and molybdenum (Mo). Sodium (Na) and iodine (I) are also required by animals.

The carbon:nitrogen (C:N) ratio is an indicator the ecologists often use to express the relative nitrogen content in the food, and/or of organisms' tissues or in the organisms themselves. Amongst all the living organisms, the amount of carbon is comparatively higher than that of nitrogen. A comparatively higher C:N ratio will be indicative of a lower nitrogen content. With a substantially wider average C:N ratio of 25:1, plants express lower amounts of nitrogen in their composition. In case of bacteria, fungi, and animals, the C:N ratio is comparatively lower (5:1 to 10:1, on average), which indicates that they are quite rich in their nitrogen content.

C:N ratios also help ecologists understand food habits, reproduction, and decomposition of the organisms (Molles 2005).

While the plants obtain carbon from the atmosphere through their stomata and all the essential nutrients from the soil, animals obtain energy as well as C and all other nutrients through food.

HERBIVORY

Herbivory seems to be the simplest case in trophic biology. However, herbivores have to encounter specific hindrances in utilizing vegetation as their source of energy and nutrients. Plants do not have nervous systems, but that doesn't mean they are any less smart in defending themselves from their enemies. The co-existence of plants with a variety of pathogens and browsing animals has made it compulsive for them to develop defense mechanisms against herbivory. The co-evolution, forced by co-existence with herbivores, has led to the development of some astounding defense mechanisms in plants. Several plants the herbivores feed on have evolved defense mechanisms to save themselves from ending up just as feed for their enemies.

A wider C:N ratio that considerably decreases the nutritive value of the plants is the first order of defense that characterizes most of the plants. Herbivores, therefore, are not satisfied with relishing on a few varieties of plants. To overcome the wider C:N ratio (some plants, like pines, have as wider C:N ratio, as 300:1), herbivores relish on a diversity of plants so that they will fulfill their protein requirements by selecting some plants with narrow C:N ratios for their nourishment.

Physical defenses such as the presence of thorns, prickles, spines, and trichome in many plants enable them to protect themselves from being fed on by animals. While the presence of such physical defenses in several plants will keep most of the animal species away from them, some animals have adapted to feed on such plants, but will not be able to extract large amounts of biomass.

High amounts of abrasive silica in grasses considerably slows down animals' feeding rates. Animals grazing on grasses with large amounts of abrasive silica are the ones that have specialized dentition.

A chemical defense with high amounts of cellulose and lignin in most of the plants would deprive the non-ruminant animals from feeding on such plants. Some animals (e.g., the ruminants) have overcome this chemical defense by depending on bacteria and protozoa in the rumen part of their digestive tract. Extracellular enzymes in the microorganisms help the ruminants break down cellulose into its glucose constituents and thus digest the same with the enzymes secreted from their own digestive system.

The presence of phenolic compounds, such as tannins, is yet another dimension of the chemical defense evolved in some of the plants. Plants with high tannin contents would decrease protein digestibility by the animals. Many animal species would refuse eating such plant species. Still, there are numerous plant species that have developed chemical defense mechanisms by producing deadly toxins and thus saving themselves from damage by consumers.

None of the defenses evolved in the plants, however, is perfect because the same is overcome by certain animal species. At slower rates, these species, in due course of time, seem to have adapted in using the plants with physical and chemical defenses. Plant–animal relationships are specific. All plants are not used as food by all animals. Nevertheless, there is no species, however equipped with strongest possible defense mechanism, which may be impervious to be used as food by any other consumer species, be they animals, insects, or pathogens. The physical and chemical defense strategies, in fact, have ensured survival of the plants, even amidst potentially destructive consumers.

CARNIVORY

Organisms specialized in carnivory feed on selected animals from which they have to derive energy and nutrients. The energy the specialized animals (predators or hunters) derive from other animals (prey) is successive from photosynthesizers (producers), via herbivores, to primary, secondary, and tertiary (top) levels of carnivores. Carnivores (predators) have to prey upon other animals (herbivores or lower-order carnivores) for their nutritional and energy needs; hence, they would choose the prey of the size that could fulfill their needs, as per their size and nutritional requirements. However, it is not also a case that an animal would choose its prey at ease without meeting any resistance.

Of course, every prey species is born with a certain degree of defense mechanism, and many prey species being masters in this art often pose a challenge (sometimes even a threat) to their predators.

Camouflage (Greek: *kryptos* = "hidden," i.e., ability of an organism not to be noticed) is one of the most common defense strategies among prey. Predators cannot search for their prey if they are unable to spot them. *Phobaeticus chani*, the longest insect species, has its coloration, long, stick-like body, shape, and size as its form of camouflage; it practices by blending in with the vegetation it crawls upon. A dead-leaf butterfly blends with dry leaves and saves itself from its enemies. Most of the caterpillars blend in with the plants they inhabit, easily evading their enemies. A bat-faced toad found in Colombia seems to have mastered resembling its surroundings, "deceiving" its predators.

Some predators also practice camouflage to hide themselves from the sight of their prey. Deceived by the camouflage, the unfortunate prey fall victim to their patient predators. The ghost mantis found in Madagascar and Africa and the wolf spider are examples predators using camouflage with the surroundings in search of prey.

Many animals use mimicry (Greek, *mimietikos* = "imitation") of each other. Some predators (e.g., butterflies, stinging wasps and bees, poisonous snakes) mimic their prey so that they can easily feed on them. Yet some prey also save themselves, mimicking their potential enemies; for example, syrphid flies mimic bees, and king snakes mimic coral snakes. The former type of mimicry performed by harmful organisms is called as Müllerian mimicry, while the latter one practiced by harmless or non-poisonous organisms is called as Batesian mimicry (Molles 2005). Mimicry happens to be an incredible evolutionary attribute for mimetic organisms and imitated ones (the models) toward improvement in their respective survival abilities.

Behavioral defense mechanisms are virtually the common characteristics of all the organisms they demonstrate when facing immediate threat from their predators. A likely victim will behave in such a way that a conspicuous "beware of me!" warning signal is passed on to its predator. The "beware of me" warning is advertised in several ways: trickery (false appearance, formidable features, like making big eyes), warning calls, screaming, scaring, fighting, banding together in groups, hiding to take refuge in burrows, flashing bright colors, hissing, spitting, urinating, playing dead, etc.

Many organisms save themselves from ending up as prey by means of some peculiar physical or anatomical features, like repellents, poisons, sharp quills, spines, or shells. Faced by grave menace, predators often deny feeding on such well-defended organisms.

Predation is usually size selective. A tiger will not like to prey upon a rat or a rabbit. It will select prey that can satiate its hunger and fulfill its nutritional needs. Since a predator has to first subdue its prey, its size should be manageable.

The predator–prey relationship in nature appears to be co-evolutionary. Poorly defended prey will be eliminated by their predators, but the well-defended ones will be spared. This situation would adversely affect the reproductive ability of the predators. Now, the fewer offspring of the hunting populations will improve their strength to subdue and kill their prey for their nutrition. These reciprocal co-evolutionary defense mechanisms, thus, serve to maintain an ecological balance.

DETRITIVORY

Detritivores or detritophages are the animals feeding on detritus. Some examples are those belonging to Polychaeta, Crustacea, Echiuroidea, and Rotifera. The detritivores subsisting on dead-plant material consume food rich in carbon and energy but poor in nitrogen. Plants are already poor in nitrogen. Detritus originating from plants is still poorer in nitrogen content, it being nearly half that in the leaves of living plants. Again, the chemical defenses of the living plants that pose challenges to the animals are also passed on to their detritus. The detritivores, thus, face a dual problem: low nitrogen and incriminating chemicals in the detritus. These two factors reduce the rate of detritus consumption by the soil-inhabiting invertebrates.

Chemical-defense mechanisms in plants and consequently in detritus seem to have evolved to slow down the rate of organic-matter decomposition to enable a buildup of a carbon pool in the soil that is vital for ecological balance.

ENERGY LIMITATIONS

Growth, functional performance, and sustainability of organisms, populations, and communities depend on their efficiency of energy utilization. Preconditions of energy limitations in an environment can cost performance as well as existence of some species competing for limited-energy amounts. Poor energy efficiencies might be perilous to individual species and their populations. The evolutionary ecologists link energy efficiencies with the natural selection of the species.

Energy available in an ecosystem is allocated to all existing species and their populations. What happens in an ecosystem/environment also happens at an individual's level. Maximization of all ecosystem functions cannot be possible with limited supplies of energy. So is the case with individual species. If more of the limited energy is allocated for reproduction, growth and/or defense functions of the species will be compromised to a certain extent. Thus, various functions an individual performs depend on the allocation of energy for the organs/systems performing those functions. The principle governing functionwise energy allocation is known as the principle of energy allocation. Limited energy resources will compel the organisms and their populations in an ecosystem to exist in a state of competition, and those exceling in the utilization of scars and limited resources will prevail in the evolutionary line.

OPTIMAL FORAGING BY ANIMALS

In an environment/ecosystem, all the feed resources might not be available plentifully. Their limitations impose a distinction in the feeding behavior of the species. They follow what is called optimal foraging to fulfill their nutrient and energy needs so that they can allocate optimum energy for effective performance, such as growth, reproduction, protection, or defense. The optimal foraging theory proposed and elaborated by MacArthur and Pianka (1966) and Emlen (1966) suggests that organisms feed following an optimization process through which they maximize their food intake. This theory predicts what, how, when, and where the "optimal foragers" should eat in the environment they inhabit. The optimal foraging theory is a branch of behavioral ecology that studies the foraging (acquisition of foods by hunting, fishing, gathering, etc.) behavior of animals in response to the environment they inhabit.

Foraging behavior depends on some factors, such as learning or adaptive change in foraging, genetics, predation, and parasitism. Foraging is of two types, viz., solitary foraging and group foraging.

In solitary foraging, individual animals of a species search, capture, handle, and consume their prey on their own. An individual has all the necessary skills, which it uses in finding, capturing, and handling its selective prey of appropriate size that will fulfill its nutrient and energy needs. Solitary foraging is a case in ecologically affluent ecosystems where food resources are enough to satiate the needs of the inhabiting individuals.

Group foraging is performed when animals search, capture, and consume their prey in a group. The foraging behavior of a single individual may not be helpful for it to subdue its prey. Then the individual needs be complemented by other members of its own species for foraging. Group foraging involves both cost and benefit for all the group members. Subduing an animal of much larger size than an individual of the predator species would satiate hunger of all the members in a group. Group foraging is also conservative in behavior because it tends to save a prey for future use and is likely to have positive ecological implications. Lions and wild dogs often get involved in group foraging. The group foraging increases foraging efficiency up to a certain size of the foragers.

OPTIMAL FORAGING BY PLANTS

It seems somewhat inconceivable that plants forage as the animals do. However, it is true. Plants cannot move from one place to another like animals. Nevertheless, they demonstrate their foraging behavior. Their foraging behavior is expressed through their growth patterns, which often vary according to environmental conditions. They orient their crowns and roots as per their energy and nutrient requirements in an environment. Thus, plants forage simultaneously in two directions: using their crowns (the part containing chlorophyll) to harvest solar energy in the upper direction and roots to absorb nutrients from the lithosphere in the lower direction. The plants, thus, forage with some limitations, as they cannot increase their foraging area by migrating from one place to another.

How will the plants allocate energy for the development of their parts to harness energy and nutrients? If they allocate a greater proportion of energy to the development of roots to forage nutrients from deeper layers of the soils, then less energy will be left for the development of their shoot system with the crown foraging energy from sunlight. Energy allocation in this way imposes functional restrictions on the photosynthetic potential of the plant. On the other hand, if a plant allocates a big chunk of energy to its shoot system, it would stay poor in foraging nutrients from the soil and suffer from nutrient deficiencies.

The environments irradiated for longer periods, but poor in soil fertility, such as in deserts, are likely to be inhabited by plants that invest larger shares of energy in their root-system development. On the other hand, there are the environments with plentiful nutrients in their soils but facing obstruction of irradiance reaching them, such as in the multilayered forest ecosystems, in which lower-story plants invest more energy in crown development to enable them to forage the required amounts of solar energy. Thus, this upper-ground and below-ground energy investment by plants in their light-poor and nutrient-poor environments, respectively, is an essential characteristic of their physiology that helps them adapt to different environments without compromising the optimal growth they strive to achieve. Here, changes in the dimension of plants' morphologies imposed by their environments manifest in changes in their physiology so that they can acquire adaptation to their diverse environments.

Elucidating interesting theories on economic analogy in resource limitations in plants, Bloom et al. (1985) and Gleeson and Tilman (1992) suggested that plants should adjust allocation so that all resources equally limit growth. In this way, plants strike a dynamic balance in which all the resources must be equally limiting to growth.

In conclusion, the sessile plants, like the motile animals, are active foragers. In their foraging, they orient their morphology to efficiently forage a limiting resource and avoid excessive acquisition of a non-limiting resource. These morphological changes, inducing physiological changes (varying due to plasticity, species, and genetic differences) for foraging the environmental resources, are precisely in the proportion vital for optimum growth and functions all plants tend to achieve.

ENERGY FLOW IN THE BIOSPHERE

There is interconversion of the bioenergy at the level of organelles. There is a set of mechanisms of rationalization of energy for the different functions enacted by different organs and systems in an organism. There is a distinctive pattern of energy allocation for different functions a plant has to perform in times of certain adversities imposed by the environment a plant grows and performs in, as we have discussed above. All the considerations at the cellular, tissue, system, or organism's level set the stage to analyze bioenergetics at the level of the biosphere.

Radiation energy input to the earth's atmosphere is equal to $1,366$ W m^{-2}, what is known as the "solar constant." A fraction of the energy is used in chloroplasts to synthesize ATP and NADPH.

ATP and NADPH, the two energy "currencies" of living organisms, lead to the reductive fixation of CO_2 into sugars through photosynthesis. In all the cells of organisms—plants as well as animals—carbohydrates synthesized during photosynthesis are used for respiration in the mito-chondria and undergo oxidative phosphorylation, generating ATP. In the cytosol of the cell, glucose is first broken down into two molecules of pyruvate. The pyruvate enters into the mitochondria via acetyl CoA, the connecting link between glycolysis and the tricarboxylic acid (TCA) cycle, and through a series in the TCA cycle, is oxidized into CO_2 and H_2O. One mole of glucose, through its complete oxidation, yields 30 moles of ATP.

The Gibbs free energy (i.e., the energy available to do work) released upon complete oxidation of glucose is 479 kJ (mol C)$^{-1}$, which for 6C glucose ($C_6H_{12}O_6$), equals to 2,874 kJ per mol $C_6H_{12}O_6$. As many as 48 kJ are necessitated for the phosphorylation of 1 mol of ADP. Therefore, the efficiency of the Gibb's free energy (ΔG) from $C_6H_{12}O_6$ to ATP is (Nobel 2009):

$$= \frac{30 \text{ mol of ATP per mol of glucose} \times 48 \text{ kJ per mol ATP}}{2{,}874 \text{ kJ per mol glucose}} \times 100$$

$$= 50\%$$

This value means that Gibb's free energy in glucose can be mobilized to generate ATP to the extent of 50%. ATP participates in crucial cellular functions like ion transport and synthesis of biomol-ecules, such as proteins, lipids, and nucleic acids. ATP is also utilized in operationalizing metabo-lism, maintenance, and growth of organisms.

What would be the fate of living organisms—plants as well as animals—if free energy in glu-cose was not regularly supplied to their cells? They would drift toward equilibrium and die (Nobel 2009). Synthesis of the complex and energetically implausible molecules (e.g., lipids, proteins, and nucleic acids) matters for the flux of energy through the biosphere. The biomolecules astonishingly encompass a considerably greater value of Gibbs free energy than an "equilibrium mixture" with the same value of various atoms. For example, as Morowitz (1979) revealed, atoms in the non-aqueous portion of the cells have Gibbs free energy values that are 26 kJ mol^{-1} greater than those of the same atoms in the aqueous cellular portion, that is, when the atoms are in the state of equilib-rium. A minimum value of Gibbs free energy corresponds to equilibrium (Figure 2.10).

The flux of solar energy, from the sun through the plants and their consumers, results in energy enrichment of the biomolecules, guaranteeing maintenance and sustenance of the biological sys-tems of the biosphere in a state far from equilibrium (Nobel 2009). This is an essential condition for the biosphere to reverberate with life.

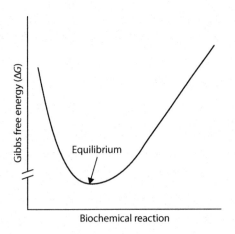

FIGURE 2.10 Gibbs free energy vs. biochemical reactions in the cells.

SUMMARY

The energy source all living organisms on Earth use is in three forms, viz. light, inorganic molecules, and organic molecules. The organisms using sunlight directly are categorized as photosynthetic (photoautotrophs) and those using inorganic molecules as chemosynthetic (chemoautotrophs), and both together are referred to as autotrophs. The organisms that use organic molecules to derive energy for themselves are called as heterotrophs. Photosynthetic autotrophs (or photoautotrophs) directly use sunlight as a source of energy and CO_2 as a source of carbon. Among all the living organisms, those belonging to prokaryotes (bacteria and archaea) enjoy living on all three types of energy sources.

Light falling on or entering into an ecosystem is not uniform in its quantity and quality. It changes with the time of a day, with weather, and with seasons at a place or in an area, or from one region to another, in accordance with latitude, altitude, and slope orientation. Even the organisms themselves are capable of inducing changes in light. In water bodies, it is only the superficial euphotic zone that receives as much light as would be able to support photosynthetic organisms.

A leaf, being a site of photosynthesis, is a key structure shaping the biosphere. Every plant species has distinctive features of its leaves, the first and the foremost feature imparting morphological distinction to a plant species. When a leaf is in equilibrium, energy input equals energy output. If it is not in equilibrium with its environment, it means the temperature of the leaf will change—increase or decrease—until equilibrium is attained. It takes less than a minute for a leaf to attain equilibrium. Short-wave and long-wave radiation are the inputs while long-wave radiation, conduction and convection, transpiration, transmission, reflectance, and energy in carbon compounds (sugars) are the outputs of a leaf in its energy budget.

The "water cost–carbon gain" is an inevitable part that photosynthesis plays. The weight of carbon gained per unit weight of water lost to the atmosphere is the water-use efficiency (WUE) = $\frac{C_{gain}\,(g)}{H_2O_{lost}\,(g)}$.

There are three different biochemical pathways of photosynthesis broadly determined according to ecological differences among the photosynthetic organisms. These are C_3, C_4, and CAM photosynthetic pathways. The organisms following these three pathways are referred to as C_3 plants, C_4 plants, and CAM plants, respectively. In C_3 plants, CO_2 is fixed following the Calvin cycle. CO_2 first combines with a five-carbon compound known as RuBP. Catalyzed by the enzyme RuBP carboxylase, the first product of the carbon-fixation reaction is a three-carbon acid, 3-PGA. It is because of the three-carbon molecules of PGA that it is called C_3 photosynthesis and the plants employing it, the C_3 plants. C_4 photosynthesis is an alternative pathway in which carbon fixation and light-dependent reactions of photosynthesis take place in separate cells. In the C_4 plants, carbon fixation in combination with PEP accomplishes in mesophyll chloroplasts, producing a four-carbon intermediate, typically malate (malic acid). This initial reaction is catalyzed by PEP carboxylase and contributes to concentrate CO_2. CAM is a sort of strategic photosynthesis adapted to extremely hot and dry environments. In this pathway, carbon fixation takes place at night when, during CO_2 uptake, water loss at lower temperatures is reduced to a minimum. CO_2 taken in combines with PEP to form 4C acids, which get stored in the large vacuoles of their photosynthetic cells. During the daytime, the stomata are tightly closed to prevent water loss, and the 4C acids are decarboxylated to release CO_2, which then enter the Calvin cycle, that is, the C_3 photosynthetic pathway. Although CAM plants do not normally perform photosynthesis at very high rates, their WUE (mass of CO_2 fixed per kg of water used) exceeds that of C_3 and C_4 plants.

Einstein's $E = mc^2$ is not valid for ecosystems. The universal sun–soil–plant connect mediated by photosynthesis is one of the most vital phenomenal outcomes of cosmic evolution—the phenomenon that imparts a living dimension to the universe. The organisms—only among the microorganisms—that are capable of deriving energy from the reactions involving inorganic molecules to synthesize complex organic compounds are the second category of autotrophs: the chemoautotrophs. This phenomenon occurring in the absence of light is known as chemosynthesis. All the heterotrophs sustain themselves by deriving energy from organic molecules. These organic molecules originally are a "product" of photosynthesis. The organic molecules in the plants used as foods emanate from the primary productivity of an ecosystem, while those in the animals constitute the secondary productivity of an ecosystem.

There are just five constituents, viz., carbon, hydrogen, oxygen, nitrogen, and phosphorus—or CHONP—that make up 93%–97% of the biomass of all organisms on Earth: bacteria, protozoa, fungi, plants, and animals. Among all five living kingdoms, plants are chemically the most distinctive, often containing lower contents of nitrogen and phosphorus.

Herbivory seems to be the simplest case in trophic biology. However, herbivores have to encounter specific hindrances in utilizing vegetation as their source of energy and nutrients. A wider C:N ratio that considerably decreases the nutritive value of the plants is the first order of defense most of the plants are characterized by. Organisms specialized in carnivory feed on selected animals from which they have to derive energy and nutrients. Camouflage and mimicry are the two characteristics carnivores have to face. Behavioral defense mechanisms are virtually the common characteristics of all the organisms they demonstrate when facing immediate threat from their predators. The predator–prey relationship in nature appears to be co-evolutionary. Detritivores, or detritophages, the animals feeding on detritus, face a dual problem: low nitrogen and incriminating chemicals in the detritus.

In an environment/ecosystem, all the feed resources might not be available plentifully. Their limitations impose a distinction in the feeding behavior of the species. They follow what is called optimal foraging to fulfill their nutrient and energy needs so that they can allocate optimum energy for effective performance, such as growth, reproduction, protection, or defense. The foraging behavior depends on some factors, like learning or adaptive change in foraging, genetics, predation, and parasitism. Plants growing in nutrient-poor soils invest more of their energy in root development, while those receiving relatively less light in an ecosystem invest largely in their canopy development.

Biological systems of the earth maintain and sustain themselves in a state far from equilibrium, a precondition set by constant energy enrichment of biomolecules through energy flux from the sun to the plants and consumers.

REFERENCES

Béjà, O., Aravind, L., Koonin, E. V., Suzuki, M. T., Hadd, A., Nguyen, L. P., Jovanovich, S. B. et al. 2000. Bacterial rhodopsin: Evidence for a new type of phototrophy in the sea. *Science* 289: 1902–1906.

Béjà, O., Spudich, E. N., Spudich, J. L., Leclerc, M. and Delong, E. F. 2001. Proteorhodopsin phototrophy in the ocean. *Nature* 411: 786–789.

Bloom, A. J., Chapin, F. S. and Mooney, H. A. 1985. Resource limitation in plants—An economic analogy. *Annu. Rev. Ecol. Evol. Syst.* 16: 363–392.

Brown, R. H. 1999. Agronomic implications of C4 photosynthesis. In Sage, R. F. and Monson, R. K. (eds.) *C4 Plant Biology*. San Diego, CA: Academic Press, pp. 473–507.

Christopher, G. and Casper, E. 2002. *Light Transmission through a Forest Canopy*. 19 p. http://physics.dickinson.edu/~sps_web/research/Forest.pdf. (Retrieved July 29, 2017).

Emlen, J. M. 1966. The role of time and energy in food preference. *Am. Nat.* 100: 611–617.

Gleeson, S. K. and Tilman, D. 1992. Plant allocation and multiple limitation hypothesis. *Am. Nat.* 139 (6): 1322–1343.

Jorgensen, S. E. 1992. Exergy and ecology. *Ecol. Model.* 63 (1–4): 185–214.

Lloyd, J. and Faquhar, G. 1994. ^{13}C discrimination during CO_2 assimilation by the terrestrial biosphere. *Oecologia* 99: 201–215.

MacArthur, R. H. and Pianka, E. R. 1966. On the optimal use of a patchy environment. *Am. Nat.* 100 (916): 603–609.

Molles, M. C. 2005. *Ecology: Concepts and Applications*. Boston, MA: McGraw Hill. 622 p.

Morowitz, H. J. 1979. *Energy Flow in Biology*. Woodbridge, ON: Ox Bow Press.

Nobel, P. S. 1991. *Physiochemical and Environmental Plant Physiology*. San Diego, CA: Academic Press.

Nobel, P. S. 2009. *Physicochemical and Environmental Plant Physiology*, 4th edn. Amsterdam, the Netherlands: Elsevier Academic Press. 582 p.

Reicosky, D. C. 2007. Carbon sequestration and environmental benefits from no-till systems. In: Goddard, T., Zoebisch, M., Gan, Y., Ellis, W., Watson, A. and Sombatpanit, S. (eds.) *No-Till Farming Systems*. Bangkok, Thailand: WASWC, pp. 43–58.

Sage, R. F. 2004. Tansley review: The evolution of C4 photosynthesis. *New Phytol*. 161: 341–370.

Sage, R. F., Li, M. R. and Monson, R. K. 1999. The taxonomic distribution of C4 photosynthesis. In: Sage, R. F. and Monson, R. K. (eds.) *C$_4$ Plant Biology*. San Diego, CA: Academic Press, pp. 551–584.

Singh, V. 2019. *Fertilizing the Universe: A New Chapter of Unfolding Evolution*. London, UK: Cambridge Scholars Publishing. 286 p.

WEBSITES

http://soilquality.org/indicators/total_organic_carbon.html.

http://www.bbc.co.uk/schools/gcsebitesize/science/add_edexcel/organism_energy/photosynthesisrev1.shtml.

https://doi.org/10.1016/0304-3800(92)90069-Q.

https://oceanexplorer.noaa.gov/explorations/06mexico/logs/may16/media/riftia_600.html.

https://ourclasspages.wikispaces.com/Giant+Tube+Worm.

https://www.google.co.in/search?q=Energy+budget+of+a+leaf&rlz.

3 Nutrient Relations

The soil is the great connector of our lives, the source and destination of all.

Wendell Berry

THE PEDOSPHERE

The pedosphere (Greek: *pedon*—"soil" or "earth," *sphaira*—"sphere") is the sphere of soil, the outermost layer of the earth that accommodates soil-formation processes and serves as a "home" to numerous living species, largely of microscopic size, but also of numerous meso- and macrosized organisms. The pedosphere encompasses part of lithosphere, hydrosphere, and atmosphere and interacts with all the components of the environment (Figure 3.1). The pedosphere is, undoubtedly, the most fertile and ecologically most-active portion of the biosphere.

The pedosphere is composed of mineral, fluid, gaseous, and biological components and serves as a mediator of all biogeochemical cycles. The sedimentary nutrient cycles are rooted in the soil. Gaseous cycles also have strong connections with the pedosphere. Lying above the hydrosphere and lithosphere, the pedosphere wears a green mantle of vegetation and constitutes what is known as the Critical Zone of the biosphere. The Critical Zone is greatly influenced by environmental factors operating over a geographical region and by anthropogenic activities.

In the pedosphere, pedogenesis (the process of soil formation) goes on continuously. Pedogenesis begins with the physical and chemical breakdown of minerals, leading to the formation of initial material overlying the bedrock substrate. Initial biological processes are not required. However, biological components (e.g., mosses, lichens, and seed-bearing plants) help quicken the soil-formation process by secreting acidic substances, mainly fulvic acids that further help in breaking the rocks apart. Numerous inorganic reactions, weathering, and decomposition products slowly make a coherent soil body in which fluids flow laterally and vertically through the soil profile, and ion formation among solid, liquid, and gaseous phases takes place. Over a long period of time, the physical, chemical, and biological processes in the soil lead to the formation of a fertile, ecologically active, and productive pedosphere.

The pedosphere harbors most of the life on Earth and is also an essence of the above-pedosphere life. It serves as a home to most of the essential nutrients the plants of the terrestrial ecosystems are nourished by. The plants also depend on the pedosphere for their water needs. The pedosphere determines the characteristics of the terrestrial ecosystems. A healthy and vibrant pedosphere bears a healthy and vibrant ecosystem.

SOIL AS AN ECOSYSTEM

Soil is not just a physical substratum to allow plants to anchor their roots. Soil is also not a mixture of several compounds. Soil is the largest ecosystem, far more complex than other ecosystem on Earth. This ecosystem is extremely enriched by all sorts of life: micro-, meso-, and macrosized organisms, autotrophs, both photosynthesizers and chemosynthesizers, heterotrophs (herbivores, carnivores, omnivores), prokaryotes, eukaryotes, multicellular organisms, monerans, protists, fungi, plants, and animals. Biodiversity in the soil ecosystem is far richer and more fascinating than that of the aboveground ecosystems.

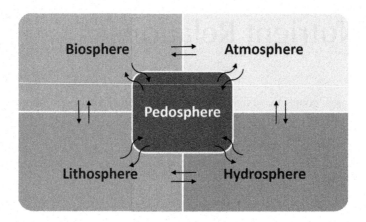

FIGURE 3.1   Pedosphere: linkages with environmental components.

In the soil environment, living organisms are in reciprocal relationships with their physical environment. Like all other ecosystems, terrestrial as well as aquatic, soil harbors its own community comprising producers, consumers, and decomposers. Food chains pervade within the soil environment. Among the food chains, detritus is dominant, but grazing food chains also occur. In fact, a complex food web operates within the soil environment. Energy and nutrient flows in the soil are akin to those in other ecosystems of the earth. Thus, ecological pyramids of the soil ecosystem might be plotted, and soil ecology can be understood.

There is enormous diversity in the world's soils and so among the soil communities. Soils change due to geographical, geologic, chemical, climatic, biological, and anthropogenic factors operating in a geographical region. These factors phenomenally influence the structure and functioning of a soil community. Biodiversity in the terrestrial ecosystems—grasslands, forests, savannas, cultivated lands, etc.—also greatly influences the soil community as well as the nutrient cycles and energy flows within the soil environment.

Traditional farmers in India hold a reverential attitude for soil. For them, soil is a living and a sacred system. According to an adage: Feed not the plants, feed the soil, so that soil itself feeds the plants. Traditional farmers trust in maintaining the health of their soil, because for them, healthy agriculture is a manifestation of a healthy soil (Singh 2016).

The biotic components of the soil ecosystem baffle human minds in several ways:

- There is maximum life (number of organisms) per unit area and per unit volume of the soil.
- Biomass of soil microorganisms alone is more than that of all human beings, elephants, and whales of the earth put together.
- A pinch of fertile soil has in it the numbers of microorganisms that are more than all human beings on the planet.
- Most of the organisms belonging to three out of five kingdoms of the living organisms, viz., Monera, Protista, and fungi inhabit the soil ecosystem.
- A teaspoon of soil from a native grassland ecosystem contains: bacteria numbering 600–800 million from as many as 10,000 species, 10,000 protozoa, 20–30 nematodes from 100 species, and several fungi kilometers long.
- The "invisible world" of the soil is more fascinating than the visible world.

DIVERSITY OF THE WORLD'S SOILS

Earth's surface is not leveled or smooth. It is full of heterogeneity. It implies that Earth's land resources are also heterogeneity ridden. Diversity of soil (pedodiversity) is the manifestation of the diverse natural features of the earth. And diversity among the terrestrial ecosystems and among species and their genotypes all over the earth is the manifestation of soil diversity. Soils of the world are varied in accordance with their geographical positioning/climatic region, physical structures, chemical composition, and biological features. Climatic or environmental factors, seasons of the year, geologic factors, and hydrologic factors are also the dominant factors that determine the physicochemical and biological properties of the soils. The vegetation cover of a soil influences the soil environment to a great extent. Because vegetation types are largely determined by humans' socioeconomic activities, as of our contemporary world, anthropogenic factors remain one of the dominant reasons to influence soil ecology.

SOIL TYPES

There are six main soil types planet Earth exhibits: sandy, silty, clay, loamy, peaty, and chalky. These soil types are made up of three major soil types, viz., sandy, silty, and clay.

Among all the soil types, sandy soil has the largest particles. Upon touching the sandy soil, one feels it to be gritty and dry. Particles in the sandy soil have huge spaces, which is why this soil type cannot hold water. Water drains through this soil type very quickly, and it is difficult for plant roots to search for water. Further, plant growth in such soils is poor because the soil, due to rapid runoff, loses much of its nutrients.

Silty soil, compared with the sandy soil, has a much smaller particle size and gives a somewhat smooth touch. Dirt would be left on the skin if it comes into contact with silty soil. This soil type is good in its water-retentive quality but is poorly aerated.

Among the three major soil types, namely sandy, silty, and clay, the clay type has the smallest particles, a feature that determines the quality of this soil. Clay soil has a very good water-retentive quality, far greater than that of the silty type. When dry, it is quite smooth, but when wet, it is sticky. Tiny particles of clay soil do not allow much air to pass through. Further, because of the slow drainage property, clay soil is capable of holding plenty of nutrients within. Clay soil, thus, is good for plant growth.

Peaty soil is black and dark brown. It gives a soft and somewhat spongy touch. It is very rich in organic matter and has a high water-retentive quality and even becomes saturated with water. The history of the peaty-soil-type formation is often traced back 9,000 years when glaciers began to melt rapidly. Vegetation that came into the way of flowing water died. Biodegradation of the water-embedded vegetation was slow, and with the passage of time, organic matter accumulated and peaty soil came into existence. It is good for plant growth. Due to its moisture-retentive quality, peaty soil supports plant growth even during a drought spell.

Chalky-type soils, also called basic soils, are alkaline in nature. The soil is soft to the touch. Soft rocks made up of chalk breakdown, easily contributing to the formation of chalky or saline soils. These soils are easily drained. The water-retention capacity is very poor. The soil dries very quickly. Chalky soils may be fertile for certain plants. However, due to high alkalinity, certain soil nutrients, for example, iron (Fe), are rendered unavailable for the plants.

The loam-type soil is "perfect" in itself because it contains the balanced proportion of all the three major soil types—that is, sand 40%, silt 40%, and clay 20%—along with humus. It is dark in color and soft, brittle, and powdery to the touch. It has a very high water-holding capacity. It preserves moisture even during a dry spell, which often is enough to sustain plant growth. The soil is often characterized with an appropriate proportion of organic matter, essential plant nutrients, and moisture. Because the loam is stocked with essential nutrients with high availability for the plants, it is considered the ideal soil type for crop cultivation.

SOIL BIODIVERSITY

Soil is undoubtedly the most complex ecosystem as well as the most diverse habitat on Earth. In no ecosystem of the world are the species more packed as in soil. The world's soils show an immense diversity in terms of the biodiversity they inhabit. Soil biodiversity varies from region to region, from area to area, from one place to another in the same area and even from one field to another field at the same place. Biodiversity within soil environments is much more complex and worth appreciating than in the above-soil terrestrial environments.

Soils host a quarter of our planet's biodiversity. A scenario of the diversity of life within the soil habitats of the world are presented in a nutshell by FAO (2015), which really baffles the human mind as follows:

- More than 1,000 species of invertebrates may be found in a single square meter of forest soils.
- The world's several terrestrial insect species dwell in soil for at least some stage of their life cycle.
- One gram of soil might contain millions of individuals and thousands of bacteria species.
- A healthy soil may contain thousands of species of bacteria and actinomycetes, hundreds of species of fungi, tens of species of nematodes, 50–100 species of insects, 20–30 species of mites, several species of earthworms, and several species of vertebrate animals.
- Soil contains organisms with the largest area; for example, a single colony of the honey fungus (*Armillaria ostoyae*) may cover approximately 9.0 km^2.

Land-use practices are the major determinants of soil biodiversity. Soils of natural forests are richer in biodiversity than those of the grasslands. Soils of the cultivated systems have different sets of biodiversity. Human intervention with the soils drastically affects structure, composition, biodiversity, organism activity, and functioning of the world's soils.

Soil organisms do not just use the diverse soils as their habitats. They play vital roles in the soil ecosystem influencing the atmospheric, hydrological, and biological systems through their interactions. They modify a soil's physical structure and chemical composition. They maintain the key services leading to nutrient cycling, gaseous exchange, soil-C sequestration and greenhouse emissions, enhancement of nutrient supplies to plants and, consequently, maintenance of plant growth and production performance. Biodegradation of organic matter by soil organisms, apart from releasing greenhouse gases, contributes to mineralization. Several organisms in the soil habitats contribute to soil detoxification. Some of them, of course, serve as foods and medicines. Soil organisms exist in mutualistic relationships and, by means of improving ecosystem functioning, ameliorate productive performance of agricultural systems. Soil organisms vitally influence the above-soil ecosystems and the inhabiting communities.

Pedodiversity–Biodiversity Relations

Pedodiversity (variety of soils) and biodiversity (variety of living organisms), according to Ibáñez and Feoli (2013), conform to the power law at the planetary level. Diversity of the soils and biodiversity strongly correlate with each other at the global level. "When a country has high pedodiversity," argue Ibáñez and Feoli, "it has also high biodiversity." Pedodiversity can be looked at as a measure of environmental heterogeneity. The biodiversity of an area, according to these workers, depends on the extent of pedodiversity of that area as also do environmental heterogeneity and the variety of habitats. Pedodiversity, thus, can reliably be used as an indicator of biodiversity.

The pedodiversity–biodiversity relationships may possibly determine the overall features and functioning of a terrestrial ecosystem. The above-ground biodiversity is rooted into pedodiversity. Pedodiversity, in fact, serves as the basis for the biodiversity to blossom on planet Earth. In measuring the earth's pedodiversity, we can map the biodiversity of the planet and also predict the evolution of natural ecosystems. In our contemporary world, most of the earth's ecosystems have been modified thanks to human interference. Most of the pedosphere, thus, is also being influenced by the human species. This anthropogenic intervention is also phenomenally influencing the vital pedodiversity–biodiversity relationships, leading to retrogressive succession of natural ecosystems.

ESSENTIAL ELEMENTS

There are 16 elements that are regarded as essential for all the plants we cultivate: carbon (C), hydrogen (H), oxygen (O), nitrogen (N), phosphorus (P), potassium (K), calcium (Ca), magnesium (Mg), sulfur (S), copper (Cu), zinc (Zn), manganese (Mn), iron (Fe), boron (B), chlorine (Cl), molybdenum (Mo). However, sodium (Na), silicon (Si), and cobalt (Co) are also essential for some plants (Gardner et al. 2018). Among these nutrients some are required in relatively large quantities, which are called major elements, or macronutrients. These include C, H, O, N, P, K, Ca, Mg, S, and Fe. Some elements, on the other hand, are required by the plants in relatively smaller quantities and are called microelements, or micronutrients or trace elements. These are Zn, Cu, Mn, B, and Mo.

A list of the nutrients that are integral constituents of certain biomolecules in plants is presented in Table 3.1. These elements are also found in organic compounds.

How is the essentiality of an element established? There are two ways to declare an element essential:

1. when the plant grown in a medium devoid of the metabolite that the element is a constituent of fails to grow or complete its normal life cycle; and
2. when the element is a constituent of a metabolite necessary for playing a physiological role (e.g., sulfur is a constituent of the amino acids cysteine and methionine).

TABLE 3.1
Nutrients as Major Biological Constituents/Biomolecules in Plants

Elements that are:	Element	Biological Constituents/Biomolecules	Sources
Components of major biological/organic compounds	C, H, O	All biological/organic matter	CO_2, H_2O
	N	Nucleic acids, proteins	NO_3^-, NH_4^+
	S	Proteins, coenzymes	HSO_4^-, HSO_3^-
	P	Nucleic acids, nucleoside triphosphates (NTPs), Phospholipids	$H_2PO_4^-$
	Ca, B, Si	Cell wall	Ca^{2+}, BO_3^{-3}, $Si(OH)_4$
Major components of coenzymes	Mg	Chlorophyll, ATP complex	
	Fe	Heme, ferridoxin	
	S	CoA, biotin, thiamine, lipoic acid	
Involved in enzyme activation	Na	C_4, CAM photosynthesis	
	K	Osmoticum, ribosomes, 40 other enzymes	
	Cl	PS oxygen evolution	
	Zn	Alcohol dehydrogenase, glutamate dehydrogenase, carbonic anhydrase, etc.	
	Cu	Ascorbate oxidase, phenoloxidase, plastocyanin	
	Mg	ATPases, kinases, RuBisCO, ribosomes	
	Mn	PS oxygen evolution, Kreb cycle	
	Mo	Xanthine dehydrogenase, N fixation, N reduction	
	Co	N fixation	
	B	Carbohydrate translocation	
	Ni	Urease	

Requirements of certain mineral elements vary from species to species. For instance, while lower forms of plants require fewer elements, the higher forms are likely to require all the 16 mineral elements previously mentioned. Ca and Mg are macronutrients for higher plant forms but serve only as micronutrients for fungi. Yet, there are certain minerals that are essential only for a few, not all, plant species. For example, Si is essential for rice, one of the major crops grown extensively throughout South Asia. Si, as it was established by Elawad et al. (1982), is an essential mineral element for sugarcane also. Na, as a microelement, is a requirement of *Halogeton*, a saline soil weed, and *Atriplex vesicera*, a pasture plant in Australia, as was shown first of all by Williams (1960) and Brownell (1965), respectively. Brownell and Crossland (1972) found that Na is a requirement of certain C_4 plants and that chlorosis and necrosis occurred in the leaves of the experimental C_4 plants when Na was withdrawn.

N-fixing organisms in soils and aquatic medium, both symbiotic and free-living, essentially require Co, which is indispensable for the synthesis of cyanocobalamin (vitamin B_{12}) by these organisms.

It is the technique of hydroponics that can readily help establish the essentiality of a specific mineral element for a plant or a crop. It is because an element to be examined can be withdrawn by dissolving chemically pure salts in deionized water. While it is quite easy to test the essentiality of an element for a plant, examining its non-essentiality is not quite a sensitive technique.

SOURCES OF PLANT NUTRIENTS

Plant-nutrient sources include inorganic and organic substances. These nutrients come from soil, water, and atmosphere—the components of the biosphere. Organic substances contain inorganic elements, which, upon biodegradation, are released into environment, becoming available for plants. More than 75% of the soil comprises Si, aluminum (Al), and oxygen (O), which do not directly serve as plant nutrients. The atmosphere is composed of 79% N, which is the sole source of N for plants. Carbon dioxide shares only 0.0415% of the atmospheric composition and is the sole source of C for plants. Nutrients in water (as soil solution) are present in the form of cations and anions.

Plant nutrients in their respective pools are maintained by means of nutrient cycles. Since green plants are autotrophs, they are powered to convert simple inorganic substances (their nutrients) in the atmosphere, water, and soil into complex organic substances using the energy in sunlight. Nutrients enter the plants as ions or elements and, through food chains, pass on to various trophic levels (herbivores to top carnivores) and eventually return to the environment as elements upon biodegradation by microorganisms. N in molecular form (N_2) and C in oxidized form (CO_2) return to their atmospheric pools while the other nutrients, the mineral elements, return to the lithosphere and hydrosphere.

The nutrient cycles go on endlessly in the biosphere and, thus, living organisms get continuously nourished while the nutrient pools get constantly recharged with nutrients. However, the respective nutrient pools are now being affected by the processes deviating from attaining equilibrium. For example, CO_2 in the atmosphere has been increasing mainly due to the indiscriminate combustion of fossil fuels. Rising from about 290 ppm in preindustrial times, the current ambient concentration of CO_2 has reached the level of 415 ppm. N and P are being increasingly diverted from soils to aquatic ecosystems. Eutrophication of water sources is exacerbating due to the nutrient enrichments of water bodies. Domestic, municipal, and industrial wastes are changing the pathways of nutrient cycles thanks to human management. Soils are becoming impoverished of essential plant nutrients and, in turn, are being increasingly polluted by heavy metals and toxic chemicals. Modern agricultural practices have diverted nutrient cycle paths to a great extent.

NUTRIENT AVAILABILITY FOR PLANTS

It is not the absolute quantity but the availability of nutrients that matters to determine the nutrient status of the plants. The primary factor changing the forms of the nutrients and determining nutrient solubility, hence their availability for the plants, is the soil pH. Generally, soil pH between 6.0 and 7.5 is considered appropriate for making most of the nutrients available for the plants. All the nutrients, however, do not become readily available at the same pH or at the same pH range. For instance, alkaline soils are appropriate for increasing Ca, Mg, Mo, and K availability for the plants. Such soils, on the other hand, reduce the availability of Zn, Mn, and B. Acidic soils increase the solubility of Mn, Al, and Fe, and these elements in excess are toxic for the plants.

Heavy N fertilization is an inevitable practice in modern Green Revolution agriculture. However, heavy doses of N in crops of the Gramineae (Poaceae) family—wheat, rice, and maize—increase acidity inducing Al toxicity, low-base saturation, and Ca, Mg, and K deficiencies (Gardner et al. 2018). Nitrification of nitrogenous fertilizers is the main cause of the acidification of agricultural soils.

When inorganic compounds are converted to organic compounds by plants or microorganisms, it causes immobility of mineral nutrients making them inaccessible for the plants. If biomass or organic matter with high C:N ratios (such as crop residues or saw dust) is added to soils, mineralization by microbial actions is reduced, which causes microbial immobilization. Immobilization of N and other nutrients by microbes leads to nutritional deficiencies in plants. Making soil free from microbes by means of sterilization induces the release of micronutrients like Mn, but the same might be toxic for the plants.

Unfavorable pH and other reasons previously mentioned make some mineral nutrients become unavailable or cause their deficiencies in plants. Mineral elements like Fe and Mn are converted into insoluble salts in the soils that are not absorbed by the plants. To overcome this problem, organic compounds containing the micronutrients, called chelates, are applied to ensure greater nutrient availability for plants. A common chelating agent used for Zn, Fe, and other micronutrients is ethylenediaminetetraacetic acid (EDTA). Ethylenediaminedi-o-hydroxyphenylacetic acid (EDDHA) is considered better than EDTA in calcareous soils. In the last century, Schatz et al. (1964) had revealed that biologically active humus in soil had the ability to chelate Pb, Hg, As, Zn, and Cu found in some biocides applied to soil and crops. The soil organic matter (SOM), thus, helps overcome or counteract certain toxic substances (or soil pollutants).

INTERACTIONS AMONG IONS

When the supply of a nutrient affects the uptake, distribution, or function of another nutrients, an interaction between nutrients takes place (Rietra et al. 2017). In soil solution, not all nutrients function independently and become available for plants in proportion of their individual quantities. Availability of an ion is influenced—enhanced or suppressed—by the presence of other ion(s). Some of the examples based on research results of several workers, as cited by Gardner et al. (2018), are as follows:

- Applications of nitrogenous fertilizers in soils results in an increase of K:Ca and K:Mg ratios.
- Fe uptake, as per the phenomenon known as the Viets Effect, may be enhanced in the presence of ions like Ca.
- Ca ions also result in enhanced uptake of K, PO_4, SO_4, Cl, bromine, and rubidium.
- Na decreases the uptake of K.
- Increased K intake decreases Ca and Mg percentage in plants.
- PO_4 may interfere with the uptake of Fe and Zn.
- NH_4^+ decreases the uptake of other cations.
- Anion uptake by the plants is generally increased with cation uptake.

TABLE 3.2

Nutrient Interaction and Antagonism

Interactive Nutrient	Soil Level of the Interactive Nutrient	Decreased Uptake/Deficiency of
Nitrogen	In the presence of high soil N levels; excess can dilute the positive effects	Micronutrients
	In low soil N levels	P, Ca, B, Fe, Zn
Phosphorus	In high P levels	Ca, Zn
	In low pH soils	B
Potassium	In high K levels	Ca, Mg, Fe, Zn, Cu, Mn
	High K level can make B toxic, to further accentuate Fe deficiency	B
Calcium	In high Ca level	B
	By liming and raising soil pH	B, Zn, Mn, Fe, Cu
Copper	In high Cu levels	Mo, Fe, Mn, Zn

Source: Fageria, N. K. et al., *Growth and Mineral Nutrition of Field Crops*, 3rd edn, CRC Press, Taylor & Francis Group, Boca Raton, FL, 2011.

High levels of N in the soil induce plant growth rates along with assistance in improving the availability of P, Ca, Zn, Fe, and B. However, excess N levels might dilute these nutrients, and low levels can reduce uptake of these nutrients (Fageria et al. 2011). Higher and/or lower soil levels of P, K, Ca, and Cu can decrease uptake and cause deficiency of many other plant nutrients (Table 3.2).

Plants being fed with a particular nutrient can also show the deficiency of the same nutrient. This occurs due to antagonism among nutrients in which an excess of a nutrient can block the absorption of another nutrient that the plant needs. Antagonism between two plant nutrients occurs due to the interaction between the elements of the same size and charge (negative or positive). Many examples are shown in Table 3.2. The most common examples of antagonism are:

- Mg blocks the absorption of Ca, and vice versa.
- Fe blocks the absorption of Mn, and vice versa.
- K blocks the absorption of Ca and Mg.

Deficiency of some nutrients might also occur due to mixing up and bonding between two (or more) elements, which make a salt that cannot be absorbed by plant roots. This often occurs when a base or acid is present in a soil solution. An example of this kind is Fe mixed with phosphate as a result of which hydrated iron phosphate ($FePO_4 \cdot 2H_2O$) is formed, which is not absorbed by plants and, thus, a plant develops deficiencies of both P and Fe.

Synergism and antagonisms among various elements can be summarized following what is called as Mulder's chart (Figure 3.2). This chart, often used by soil and plant scientists, presents in a nutshell how some soil nutrients influence the availability or uptake of each other.

The sum of equivalents of cations in a soil solution are equal to the sum of soluble anions. In this way, the amount of cations is maintained by a synergistic effect of anions. In the process of plant nutrition uptake, a cation in its overoptimal concentration acts antagonistically to the effect of other cations (Jakobsen 2009).

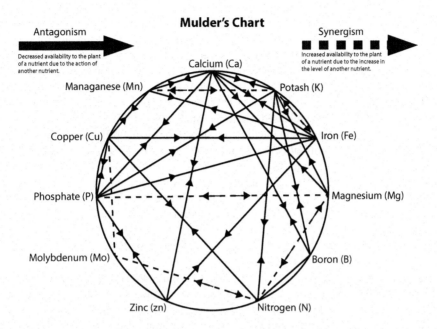

FIGURE 3.2 Mulder's chart showing synergistic and antagonistic interactions among plant nutrients in the soil. (From https://www.thedailygarden.us/garden-word-of-the-day/mulders-chart.)

CYCLES IN THE SOIL

Soil continuously supplies ions to the plants due to vegetation flourishing above the soil and, thus, a variety of communities of a variety of organisms are sustained on Earth. However, mineral nutrients do not occur in the soils in a fixed amount. They are dynamically added to or removed from the soil but tend to maintain an equilibrium, which is influenced by a number of environmental factors, both biotic and abiotic. They persist in cycles we call nutrient cycles, or biogeochemical cycles, if the mineral cycles are considered on a very large scale or in respect of the entire planet. Natural supplies of ions to the plants, therefore, are viewed in terms of nutrient cycles.

The status of ion supplies from soil to plants, viewing in terms of nutrient cycling, is especially valuable for N. Supplied to the plants as NO_3^-, N deficiency often becomes critical for the food crops. N is naturally added to the soils through physical (during rainfall) and biological fixation and losses by means of denitrification, leaching, and soil erosion. The total amount of N in the soil at a particular point of time is determined by the natural N cycle that varies from place to place, depending on soil structure and composition as influenced by environmental factors. The environmental factors are greatly altered by human management systems. For example, in cultivated soils, N deficiency for crop plants is often encountered, especially because of cropping intensity and crop rotations. This deficiency is overcome by applying artificially fixed N (through nitrogenous fertilizers, such as urea). External inputs of N often results in an excess of nitrate ions in the soils—through leaching and surface water runoff—groundwater, and surface-water resources.

N cycles in the soil, with the N_2 pool in the atmosphere, is a complex nutrient cycle. The organic component of the soil is of key significance. *Nitrobacter*, *Nitrococus*, and *Nitrosomonas* are the important bacteria responsible for the conversion of NH_4^+ to NO_3^-. Most of the plants, however, can take up both NH_4^+ and NO_3^-. The decomposers (a variety of fungi and bacteria) prospering on organic matter in the soil help produce NH_4^+. The decomposers, in order to carry on their metabolism and maintain structure, also require N. Should the organic matter in the soil be poor in N content, the decomposers will incorporate the limited N into their biomass, which is necessary for fulfilling their minimum N requirement. Thus, N will not be liberated, and soil, and subsequently the plants, will suffer from N deficiency. Narrowing down the C:N ratio in the leaf litter would, however, help the decomposers release NH_4^+ in the soil. When the C:N ratios of the litter range between 30:1 and 20:1, the corresponding percentage values of N are 1.2% and 1.8%, respectively. If the N values are less than this, low or no NH_4^+ will be released, creating a state of the soil fertility not healthy enough for the plants. The larger the C:N ratios in the litter the slower the availability of N to the plants. In other words, the slower the availability of N to the plants the larger the C:N ratios of the litter.

One more problem of NH_4^+ release in the soils is encountered because the leaves of the plants before leaf-fall recover most of the nutrients, resulting in further increase in C:N ratios in the leaf litter. In extreme cases, the plants before leaf-fall might short-circuit the N cycle by resorbing almost all N in the perennating organs, which store enough nutrients to take care of the organisms during an unfavorable season. The fresh litter may have a C:N ratio as high as 100:1 (Fitter and Hay 2002). Soil fertility promotes the decomposition rate of nutrient-poor, but not of nutrient-rich litter through N transfer (Bananomi et al. 2016).

Apart from N, many other minerals also recycle through SOM. A variety of minerals are there in the freshly fallen leaves. While most other nutrients are likely to be resorbed from the senescing leaves, Ca, due to its immobility within the plant, stays in the leaf litter. K is released into the soil and enters the inorganic nutrient cycle. P occurs in the leaf litter in organic form. However, in order to become available for the plants, it is released in its inorganic ion form, $H_2PO_4^-$; after that, it is precipitated or absorbed by the Ca, Fe, and Al surfaces in the soil.

ION-UPTAKE PHYSIOLOGY

KINETICS

A number of soil and plant factors are vital in regulating ion-uptake physiology. Bulk soil solution, mineralization, mass flow, and diffusion are undisputedly the overriding factors. Root physiological-uptake capacity is the major determinant of ion uptake by the plants (BassiriRad 2000).

In a plant, it is the root system that plays a crucial, critical, and key role in ion uptake. In order to be absorbed, nutrients in the soil medium must come in contact with roots. Plant roots also exercise their behavior (i.e., growth pattern) in response to the available (and non-available) nutrients. The physical and chemical relationships between a living root and soil nutrients are established through (Gardner et al. 2018):

1. Contact exchange;
2. Exchange of ions in the soil solution with H^+ in the mucigel, a slimy substance covering the root cap;
3. Ion diffusion down the chemical gradient;
4. Mass flow of the ions down the moisture gradient; and
5. Extension of the root into the ion medium.

The extension of a plant's root system through newly developed root-absorptive tissues (especially the root-hair zone) derives an advantage from unexploited soil medium quite rich in ions. This type of plant-root behavior finds new opportunities of nourishment of the plants in an efficient manner. Soil applications of P were found to increase P uptake as well as crop yield, root length, fineness, and density (Aboulroos and Nielsen 1979). Either the P concentration in soil solution or an increase in root length, or both, might be responsible for enhanced P intake by plants. Nutrient mass flow, that is, moving with water, is the primary process of nutrient uptake in some species under certain circumstances. A mass flow of K ions is likely to predominate in some soils, for example, coarse-textured Spodosols, Entisols, and Ultisols. This, however, is less prevalent in Mollisol. Because of the immobility of Ca in the plant, root extension is especially beneficial for this nutrient, while the contribution of this process for all other nutrients is very little.

The root cells of a plant contain much higher concentrations of ions than the surrounding medium. The root cells are also electrically negative in comparison to their surrounding soil solution. The transport of H^+ across the membrane by means of proton pumps, that is ATPases, creates this potential in the range of -60 to -250 mV. Involvement of ATPases implies that energy is expended for the maintenance of potential difference across the membrane. Most cations in the surrounding medium, as a consequence, enter the cell passively. The passive transport of an ion is down the electrochemical gradient but usually against the chemical concentration gradient for that ion. Cations, in exchange of protons in some of the cases, especially for Na^+, actively export back to the soil solution. In the company of the protons, anions are actively transported to root cells against the electrochemical gradient, and, in most of the cases, against concentration gradient (Fitter and Hay 2002). ATP energetics involved in active absorption causes the ions to cross the cytoplasm membrane (the plasmalemma).

Intercellular transport of ions in plant roots takes place through the plasmodesmata, the living connectors between the cells, and this is called the symplast pathway. Occurring through the concentration gradient, the ions passively diffuse through the cells of the cortex and endodermis to be eventually delivered into the xylem vessel or tracheids. The movement of the ions from cell to cell across their plasma membrane actively and passively through the cortex and epidermis cells is called the transmembrane pathway. Furthermore, the movement of ions through cell walls and intercellular spaces between the cells in the cortex is called the apoplast pathway. Upon reaching the innermost cortex cells, the Casparian strips on the lateral wall of endodermal cells retard further movement of the ions. The ions, therefore, enter the special cell known as the transfer cell. From here, the ions are actively taken into the xylem vessel or tracheids. The xylem of the root, shoot, and leaf, eventually, facilitates the upward movement of the ions along with transpiration stream.

Movement of the ions within root cells is quite sensitive to a number of factors, dominantly the ones shown in Figure 3.3. Each factor is not independent in itself. They are all interdependent and contribute to sensitizing the process of ion movement. These interdependent factors are constituents of an energy-dependent system facilitating ion movement across the cell membrane.

The rate of root uptake of the ions increases with an increase in ion concentration. But with a continuous increase when the kinetics saturation is reached, the ion uptake becomes independent of

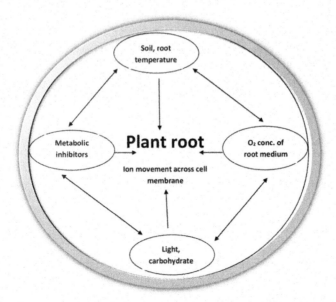

FIGURE 3.3 Factors influencing ion movement across cell membrane in plant root.

the ion concentration. Ion transporters are the membrane-bound proteins with high degrees of ion specificity. The mechanism is akin to substrate-enzyme action. The carrier-mediated transport of the ions across the root system of the plant, thus, can be described with the help of the Michaelis–Menten equation (Figure 3.4), Hofstee plot (Figure 3.5), and Lineweaver–Burke plot (Figure 3.6):

$$v = \left(V_{max} \times C_{ext}\right)/\left(K_m + C_{ext}\right)$$

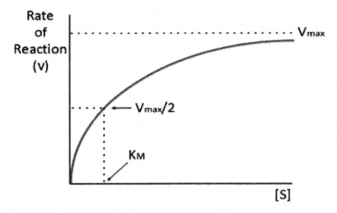

FIGURE 3.4 Absorption isotherm for K^+ ions (hypothetical data). (Based on Fitter, A.H. and Hay, R.K.M., *Environmental Physiology of Plants*, 3rd edn, Academic Press, London, UK, 367p, 2012.)

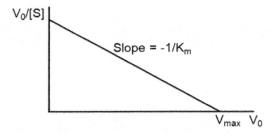

FIGURE 3.5 Hofstee plot. (Based on Fitter, A.H. and Hay, R.K.M., *Environmental Physiology of Plants*, 3rd edn, Academic Press, London, UK, 367p, 2012.)

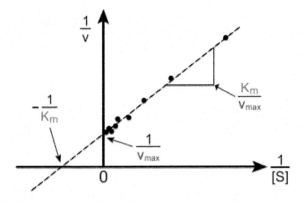

FIGURE 3.6 Lineweaver–Burke plot. (From https://en.wikipedia.org/wiki/Lineweaver–Burk_plot.)

where v is influx and K_m (Michaelis constant) represents external ion concentration (C_{ext}) at which $v = 1/2\ V_{max}$. K_m, therefore, describes a property we may refer to as "affinity" of the uptake system. The low K_m value implies high affinity.

Several workers (e.g., Van den Honert and Hooymans 1955, Becking 1956, Epstein 1976, Behl et al. 1988, Pilbeam and Kirsby 1990) elucidated that the Michaelis–Menten model can describe the net uptake of most of the essential ions, including NO_3^- and NH_4^+.

The cation-exchange capacity of the root-influencing nutrient uptake varies with plant species, variety (genotype) of the species, and age of the plants.

The vacuole of the cell plays a crucial role in striking a balance for stabilizing the cellular demand and supply of ions.

A number of environmental factors and internal factors influence ion transports, both passive and active, between the soil solution and plant roots. Ionic-substrate concentration and temperature are the major factors influencing the ion-uptake processes. With energy supplies in adequate amounts, the concentration gradient becomes the dominant factor to influence ion uptake by the plants (BassiriRad 2000).

The importance of a high-respiration rate in inducing active absorption is obvious because it generates high-energy phosphate bonds (ATP). Anaerobic conditions, thus, reduce ion uptake. Although the situation would vary according to species and ecological conditions, the low temperatures inhibit absorption while high temperatures (up to the metabolic requirement) help enhance ion uptake.

How the environmental factors altering in response to climate change, especially the increasing atmospheric CO_2 levels and elevated soil temperatures, are set to influence nutrient uptake are depicted in Figure 3.7. In the era of global warming, an increase of CO_2 concentration in the atmosphere is also influencing nutrient uptake by the plants. It is because CO_2 induces changes in the morphology and growth pattern of roots (Tingey et al. 2000) as well as mycorrhizal association (Treseder and Allen 2000). These CO_2-induced changes should affect nutrient-uptake kinetics.

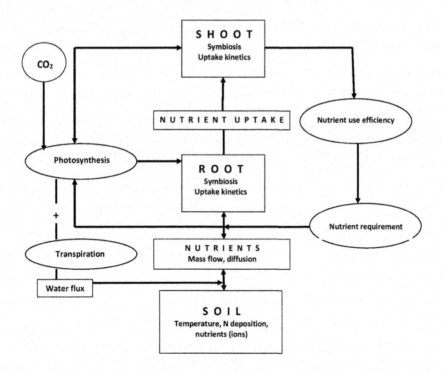

FIGURE 3.7 Nutrient uptake as affected by environmental factors induced by climate change.

Elevated CO_2 levels influence photosynthesis, which, in turn, enhance the growth rate of the plant. An increase in the growth rate increases the plant's nutrient demand along with an increase in uptake capacity. According to yet another model based on whole-plant C-nutrient balance elaborated by some workers, notably by Bloom et al. (1985), Johnson (1985), and Gutschick and Kay (1995), resources of abundant availability (i.e., CO_2 for C fixation or photosynthesis) must be allocated to increase the uptake of the most limiting resources (i.e., nutrients in the soil).

Global-warming-led climate change is bound to alter "physiology" of the whole biosphere. Temperatures of the earth are soaring. The climate models constructed by climate scientists project a further rise in global surface temperature of 0.3 to 1.7°C (0.5 to 3.1°F) to 2.6 to 4.8°C (4.7 to 8.6°F) during the twenty-first century subject to the rate of greenhouse-gas emissions (IPCC 2013). This implies a considerable increase in soil temperature. Such a magnitude of temperature will have a significant influence on soil biology, including overall root ecology and root physiology. In some ecosystems, root-zone temperature, according to Van Cleve et al. (1981) and Risser (1985), is a key factor determining primary productivity. Geochemical processes regulating N and water availability in the soil are some of the striking changes brought about by changes in the soil thermal regime. These changes are vital in influencing root ecology, including root growth and morphology, respiration, and nutrient supplies to the plants.

INTERNAL CONTROL

The Michaelis–Menten model can characterize the kinetics of ion transport into roots. The most common response is for V_{max} to increase and/or K_m to decline, and this pattern is variable as has been reported by several researchers (Fitter and Hay 2002).

Should the value of K_m fall or of V_{max} rise, the ion uptake rate even from very low external concentrations will be increased. This has been established by several experiments on barley in relation to K^+, H_2PO_4, SO_4^{2-}, and NO_3^- uptake. According to Drew et al. (1984), after a one-day K deprivation, 14-day-old plants recorded a decline in K_m from 53 to 11 μm, that is, about a five-fold reduction in the K_m values. Some other species, such as *Arabidopsis*, have recorded similar effects (Doddema and Telkamp 1979); however, in maize, the K^+ uptake is inversely related to the K concentration (from 20 to 100 mg kg^{-1}) (Claassen and Barber 1977). Thus, a number of plants can respond to short-term fluctuations in the root surface ion concentration (i.e., ion availability) by such kinetic behavior.

Plants take up most of the nutrients in a single ionic form. Uptake kinetics is the same: depending upon pH, the phosphate is taken up as $H_2PO_4^-$ or as HPO_4^{2-}. Unlike other mineral nutrients, there are no substantial soil reserves of N. N occurs in two utilizable and interconvertible forms: NO_3^- and NH_4^+. The conversion is biological (by bacteria). The relative abundance of the two utilizable forms in the soil depends on: (1) the rates of NH_4^+ production from SOM; (2) the rates of biological conversion of NO_3^-; (3) NH_4^+ adsorption rates by soil exchange sites; and (4) the rates of NO_3^- uptake and loss through leaching and denitrification. Thus, N availability for plants varies in time and space.

When NH_4^+ is picked up by a plant, loss of protons takes place leading to acidification. The reverse is the case when NO_3^- is taken up by the plant. NH_4^+ taken up by the plant gets directly incorporated, via glutamate and glutamine, into organic compounds. NO_3^- ions do not directly become part of the organic compounds. To do that, NO_3^- first has to be reduced to NH_4^+. Two enzymes are involved in the reduction process: Nitrate reductase reduces NO_3^- into NO_2^-, and nitrite reductase reduces NO_2^- into NH_4^+.

Increased atmospheric CO_2 levels have also been linked with NO_3^- and NH_4^+ uptake capacity and growth of the plants. Data has been compiled by BassiriRad (2000) (Table 3.3).

The elevated CO_2 response to the NH_4^+ and NO_3^- and whole-plant growth is species specific. While the root capacity for NH_4^+ and NO_3^- uptake capacity in most of the experimental species recorded a decline, whole-plant growth registered an increase in almost all the plant species. The positive response to growth, however, might be a short-term response, until the counteracting factors of global warming reveal their pronounced effect.

TABLE 3.3

Response of Root NH_4^+ and NO_3^- Uptake Capacity and Growth to Elevated CO_2

Plant Species	Uptake Rates		Growth
	NO_3^-	NH_4^+	
Acer rubrum	—	↑	x
Bromus hordeaceus	↓	—	x
Lasthenia californica	↓	—	x
Plantago erecta	↓	—	x
Ceratonia siliqua	—	—	↑
Pinus taeda	—	↓	↑
P. panderosa	x	↓	↑
Bouteloua eriopoda	↑	x	↑
Prosopis glandulosa	—	x	↑
Larrea tridentata	↓	x	↑
Populus tremuloides	—	x	↑

Source: BassiriRad, H., *New Phytol.*, 147, 155–169, 2000.

↑ = increase; ↓ = decrease; — = no change; x = not determined

MORPHOLOGICAL RESPONSES

Roots serve in the uptake of soil, water, and nutrients and provide C input for soil C sequestration (Sainju et al. 2017). Transport of the nutrients through soil, not across root, is the limiting factor for the uptake of most of the nutrients. As a result, a change in ion uptake kinetics will have no, or little, bearing on ion uptake. Under such circumstances, the plant would resort to making alterations in its root morphology, which, in turn, would serve the plant in a more effective way.

ROOT:SHOOT RATIO AND ROOT:WEIGHT RATIO

Roots and shoots are functionally interdependent. Dry weights of roots and shoots in a plant are maintained within a certain balance and happen to the characteristic of a species. Root:shoot ratio (W_R/W_S) or the root:weight ratio (W_R/W_{RS}) expresses the simplest index of allocation of nutrients to plant roots. While W_R/W_S is indicative of nutrient proportions allocated between the root and shoot, W_R/W_{RS} indicates the proportion of the total plant biomass allocated to the root. Both the indicators express plasticity and are generally found increasing with: (1) low soil water supply; (2) low soil temperature; (3) low soil-nutrient supplies; and (4) low soil O.

The wider W_R/W_S and W_R/W_{RS} ratios explain the fact that when the soils are poor in moisture, nutrients, and other vital factors, a plant tends to invest more energy for root development. These ratios are likely to be narrow in the environments with fertile soils and in plant species with high growth rates. Plants growing in stress-ridden environments or environmentally adverse conditions are most likely to allocate a large proportion of their photosynthate to subterranean storage organs.

Root Diameter and Root Hairs

The root system of a plant is naturally equipped with nutrient uptake ability. All plant families, barring a few like Alliaceae, possess very specialized features behind the root tip: the root hairs. The root hairs enhance the nutrient-uptake capacity of the roots by increasing the root area within the soil for efficient exploitation of the soil nutrients. The effect of the roots due to root hairs, however, is not so crucial for uptake of mobile nutrients (e.g., NO_3^-). They accelerate a root's uptake efficiency for certain nutrients, like phosphates. Root hairs contribute to root penetration, root-soil contact, and phosphorus acquisition. They help increase phosphorus intake in fertile soils, especially in low-strength soils (Halina et al. 2013).

Soil moisture within the soil environment is not uniform. It varies greatly even at short distances. Where there is no sufficient moisture, root contact with nutrient solution in the soil may break down. A root without root hairs cannot penetrate small pores in the soil environment. Root hairs, also known as absorbent hairs, under such circumstances, play a crucial role in penetrating very small pores even of less than 5 μm in size and secrete mucilage to continuously maintain what can be called as "liquid junction" between soil and root. With no root hairs in a plant root, water and mineral uptake would occur unceasingly. Root hairs, thus, create conditions for harnessing soil resources and enhance nutrient and water uptake conducive to a plant's potential growth. Root hairs are also involved in the formation of root nodules in leguminous plants vital for biological N fixation and ameliorate soil ecology.

Root diameter is also of important consideration in nutrient uptake by plants. It is particularly an important factor when nutrient concentration in the root free space is high. Conditioned by these operating factors, the uptake occurs not just in the epidermis but also in all the cortical cells. Thus, root volume, rather than the root surface, is a more important criterion for nutrient uptake by the plants.

Root Density and Distribution

Root density (or root-length density) is an important aspect of understanding root patterns of different plants in influencing water and nutrient uptake. A root has scope to grow in and explore the unlimited volume of soil. Soil factors, therefore, become the main factors to affect mineral uptake by a plant. However, the balance is maintained by the nutrient demand of an individual plant or a plant species. The phosphate-depletion zone around the adjacent roots, however, rapidly overlaps. This results in interference of one root with the other for a supply pattern of the ions, giving way to intensive competition for nutrients between the roots within the soil environment. Such a state of the soil environment leads to an attenuation of the diffusion gradient. There is more likelihood, however, that because of the narrow depletion zone most likely not to overlap, the competition would be at high root densities for ions of low-diffusion coefficients, like those of phosphates. Increased root density serves to accelerate ion uptake of immobile, not of mobile, ions.

Interactions among interspecies (or intraspecies) roots result in reducing ion-uptake rates for all the interacting roots. Competition ability of ion uptake varies from species to species. In cereals, growth of the main root axes is independent of ion concentration. So, it continues to grow even in nutrient-poor soils, but lateral root growth occurs in the nutrient-rich zone of the soil. It can be inferred that the roots exploit fertile soil volume but are actually adapted to exploit nutrient-poor soils.

Variations in ion concentrations in soil are decisive for root growth and distribution patterns. The maximum root density will occur in fertile soil. Immobile nutrient-uptake competition in interacting roots dependent largely on lateral roots will be more intensive in nutrient-rich soil or soil zones. Since the nature of competition is species specific, the poorly competing plant species will rely on the exploitation of soil volumes with poor phosphates where root densities tend to be lower.

Root-system architecture (RSA), the spatial configuration of a root system in the soil environment, highly varies from species to species and from variety to variety of the same species. RSA includes distribution, size, and angles of the branches occupying soil volumes. For a plant, RSA provides ample opportunities to exploit soil volumes varying with nutrient strata. It tends to optimize water and nutrient uptake. The branching configuration of root branches with herringbone-like patterns are considered to be more efficient from the viewpoint of explored soil volume, whereas diffuse-branched patterns happen to be more transport efficient. The herringbone-like root patterns are most likely to prevail in the soils in which nutrient acquisition by plants limits their growth.

SOIL MICROORGANISMS

RHIZOSPHERE

The pedosphere, as we have discussed previously in this chapter, harbors remarkably large populations of bacteria, protozoa, fungi, and small invertebrates. And, of course, a number of vertebrate animals, including some mammals, also inhabit soil ecosystems. Here our focus is on microorganisms, which are pivotal in mineral nutrition of the plants and have enormous implications on soil–plant relationships. Supplies of many nutrients to plants depend on microbial decomposition of the SOM. Soil microfauna facilitates microbial degradation by breaking down the litter into small fragments and thus enabling the microorganisms to accomplish the process of mineralization. The soil microorganisms are not distributed uniformly in the soil environment. They rather aggregate around energy sources—around SOM or living root tissues. Roots of some plants especially have very close relationships with some kinds of microorganisms. They serve as living sources of energy for soil microbes. Plant roots attract most of the soil microbes and also phenomenally influence microflora in the soil ecosystem. The sphere of the soil allowing microbes to flourish in the proximity of plant roots was referred to as the "rhizosphere" by German scientist Lorenz Hiltner in 1904, although a study of the roots' influences on soil microflora was first of all systematically carried out by Starkey (1929). The rhizosphere is a unique niche for the soil microbes to colonize profusely and contribute to soil-fertility enrichment and has been an interesting subject of research in agriculture aimed at ameliorating crop productivity.

Katznelson (1946) suggested the R:S ratio for a better understanding of the degree or the extent of plant roots' effects on soil microorganisms. The R:S ratio is the ratio between the microbial population per unit weight of rhizospheres to the microbial population per unit weight of adjacent soil. Clark (1949) introduced yet another closely relating term "rhizoplane" to denote the root surface along with closely adhering soil particles.

Typical R:S ratios have been found ranging from 2 to 100, although values even as high as 2000 have also been observed (Fitter and Hay 2002). The R:S ratio in the plants can also be used for phytoremediation of different effluents from industries. On the basis of their studies on wetland plants, Sukumaran et al. (2015) suggest that the floating plants have lower R:S ratios compared to the emergent plants and that the latter with high R:S ratios (and increased biomass rate) can be used as effective sources for phytoremediation of all types of industrial effluents.

The rhizosphere is an energy-rich soil "niche" that helps numerous microorganisms to perpetuate and add to soil fertility conducive to plant growth and productivity. Continuous enrichment of the rhizosphere with energy-rich nutrients is attributable to a number of sources emanating from the plant-root system, viz.: (1) sloughing of root-cap cells of the roots in the rhizosphere, (2) death of the root cells, (3) production of mucigel by the root caps, and (4) secretion of compounds from intact root cells. Sustenance of soil microbes and consequent soil fertility status, however, will largely vary according to plant species and their genotypes, soil structure, and composition, and intrinsic edaphic factors and operating environmental factors, including variations in climatic factors, and these will be spatial and temporal variations.

Impact on Nutrient Uptake

The microbial populations prospering in the rhizosphere phenomenally influence nutrient uptake by plant roots in the following ways:

1. Altering the supply at the root surface (rhizoplane)
2. Inhibition or stimulation of nutrient uptake (i.e., interference with nutrient uptake)
3. Altering root or shoot growth (or direct root damaging)
4. Dissolution of insoluble ions and/or mineralization of organic matter

There are conflicting reports about how the nutrient-uptake rates are increased thanks to the rhizospheric microbes. It clearly appears that nutrient-uptake enhancement in the rhizosphere is due to synergistic actions of both microorganisms and plant roots.

Nitrogen Fixation

N fixation provides several soil prokaryotes with an ecological advantage but places significant constraints on N fixers (Inomura et al. 2017). Several free-living and asymbiotic bacteria in the rhizosphere play the role of N fixation. Organic matter derivable from a plant-root system serves as an energy source for some bacterial species, for example, *Azotobacter*, *Klebsiella*, and *Clostridium* to fix molecular N. The endorhizophere is the zone of the rhizosphere that comes into being due to dead cortical cells of the root. The *Azospirillum* bacterium uses these cortical cells as an energy source to fix N.

N fixation requires an input of energy. And the energy costs of molecular N fixation are pretty high. According to the estimates of Giller and Day (1985), the energy-cost values range from 4 to 174 g C g^{-1}. The direct cost of reducing N is the reason for the low-growth rate or nitrogen-fixing prokaryotes. The nitrogenase enzyme catalyzes the N-fixing process using 8 electrons and 16 ATP molecules in order to reduce each N_2 molecule to 2 NH_3 molecules (Sohm et al. 2011; Inomura et al. 2017).

$$N_2 + 8e^- + 8H^+ + 16ATP + 16H_2O \rightarrow 2NH_3 + H_2 + 16ADP + 16Pi$$

There is, thus, not enough energy for the free-living and asymbiotic microbes in the rhizosphere to fix N enough for appropriate growth of some food crops, for example, cereal crops. Only a small proportion of N could be made available for the crop, which happens to be far below the actual requirement to realize productive potential. Therefore, farmers need to apply synthetic nitrogenous fertilizers, such as urea, to meet the N requirements of cereal crops. In anthropogenic ecosystems (e.g., cultivated lands) raised with cereal monocrops, N deficiency often persists. However, in natural ecosystems (e.g., natural forests), the N-fixation role is played by symbiotic bacteria and photosynthetic-cum-nitrogen-fixing green algae (cyanobacteria), and both types of organisms have their own energy supply systems on sustained basis. Natural ecosystems, thus, hardly suffer from N inadequacies.

Two most common genera, viz., *Anabaena* and *Nostoc* belonging to blue-green algae (Cyanobacteria) are involved in asymbiotic N fixation. However, since cyanobacteria are photosynthesizers, they derive energy from their own sources. N fixation by cyanobacteria is often more prevalent in the flooded soils in which the bacteria grow profusely.

In symbiotic N fixation, bacteria in the nodules of host roots harp on direct-energy supplies from the host plant. There is only a single family of the Kingdom Plantae, the Fabaceae or Leguminosae (also popular as legume, pea, or bean family) that bears nodules harboring N-fixing bacteria. And there is only Rhizobia bacterium in the Kingdom Monera or Prokaryotae that fixes N in symbiotic relationships with legumes. Nodule formation in the roots of legumes in response to the presence of Rhizobia is a key natural process to enhance N fixation to the amelioration of soil ecology and subsequent enhancement in nature's primary productivity. Since the host is a photosynthesizing organism, a fraction of the C fixed through photosynthesis is transported to nodules in the root. The nodule itself serves as an important sink for fixed C, which becomes an energy source to drive N fixation. Energy supplies from the host to the symbiotic N-fixing bacteria on a sustained basis sustain the process of N fixation.

The other nodule forming and symbiotic N-fixing bacteria include Actinomycetes (e.g., *Frankia*), which are associated with woody angiosperms, and *Alnus* being the well-known host plant. Blue-green algae establish symbiotic relationships with gymnosperms by forming nodules on the surface roots of gymnosperm species. Light is required in the process. In the humid tropics, some free-living and many other bacteria produce nodules on the leaves of woody species.

Nodulation of the following colonization of the nodulating bacteria *Rhizobium* is an interesting process that occurs as follows (Gardner et al. 2018):

1. Root-hair deformation possibly due to the response to indoleacetic acid (IAA); IAA production is thanks to the stimulation by the bacterium (or as a response to IAA-stimulated ethylene);
2. Transference of bacterial cells as a result of the formation of an infection thread;
3. Release of the bacteria into the root cortex;
4. Formation of the nodule meristem followed by nodule expansion by cortex-cell division;
5. Enlargement of the bacteria-infected cells inside the nodules; and
6. Loss of envelop of nodule bacteria (or bacteroid) and nitrogenase with the commencement of senescence of older nodules.

Cortex, meristem, vascular system, and bacteroid zones are the four main components of a nodule. The vascular system enacts the transportation of water, sugars, and minerals to bacteroids for their nourishment following the removal of the fixed N as amino acids, amides, and allantoin (ureides). Additionally, the nodule in plant roots serves to build up a favorable environment for the N-fixing bacteria.

Some bacteria are non-nodulating and do N-fixing symbiotically but in an association with some plant species. Blue-green algae make associations with *Azolla* and lichens. Some bacteria, like Azospirillum, Azotobacter, and Spirillum are associated with C_4 grasses in tropical and semitropical pastures.

Wherever the soils are N deficient, many legumes invade the site as pioneer species. However, in the N-rich soils, competitive advantage derived from N fixation vanishes. Such ecological conditions help spare larger chunks of C to be used for plant growth rather than for N fixation by bacteria.

Influence of Environmental Factors

N fixation is a function of the soil ecosystem and is not neutral to environmental factors. A number of operating environmental factors influence N fixation, as follows:

Carbon dioxide: Environment of the N-fixing bacteria is generally quite richer in CO_2 (normally 10–100 times more) and quite poorer in O_2 than the values of these gases in the ambient air. Higher levels of CO_2 (up to 4%) are known to enhance N fixation, although additional CO_2 requirement might be ruled out under the conditions favorable for root growth (Gardner et al. 2018).

C:N ratios: Lower C:N ratios in the soil reduce nodulation and/ or nitrogenase activity, therefore resulting in reduced N fixation. It is particularly because higher soil N content in relation to Soil Organic Carbon (SOC) represses *nif* expression. In leguminous plants, NH_3 has a negative effect on N fixation. NH_3 plays an inhibitory role in N fixation by free-living bacteria as well.

Weather factors: Heat and drought spells might wipe out or decrease populations of the N-fixing bacteria, resulting in a drastic decline in N fixation. Cold weather (5°C) also reduces N fixation in legumes. High soil moisture (up to 75%) is appropriate for an enhanced state of N fixation. Root nodules have approximately 80% moisture content. High soil moisture favors both the microbial population and nodule formation, and, therefore, N fixation.

pH: One of the crucial factors influencing N fixation directly as well as indirectly is the pH of soil or of a medium plants grow in. Acidic soils do not support *Rhizobium*. Nodules formed in the low pH soils support only ineffective strains of the N-fixing bacteria. Higher soil pH is supportive of nodule formation in the plant roots.

Mineral nutrients: Mineral requirements of the N-fixing organisms are akin to those of the plants. Fe, Mo, and S are the mineral components of nitrogenase. The soils with adequate amounts of these mineral nutrients, therefore, will have a higher status of N fixation due to thriving populations of the N-fixing organisms. Soil fertilization with K has a positive influence on nodulation, nitrogenase activity, and other supportive enzymes. P responds positively to the growth of the plants belonging to Leguminosae and, thus, to N fixation. Because of its role in cytochrome system and oxidative respiration, Cu is found necessary for nodulation and, thus, for enhanced N fixation. Ca is especially essential for the growth of plant and nodule meristem, thus playing its own role in N fixation by stimulating symbiosis between the host plant and N-fixing bacteria.

Pesticides: Pesticide applications are extremely detrimental for soil flora and fauna. These chemicals do not make any discrimination between good flora and bad flora and can wipe out even N-fixing bacteria in the soils. Hg-based fungicides used for seed treatment are detrimental for N fixers. Pesticides also decrease root nodulation in leguminous plants.

MYCORRHIZAS

The most enhanced area of a plant root for the microbial population and microbial activity is the root surface, the rhizoplane. A highly specialized development of a rhizoplane association is that of mycorrhizas. This association between the fungus and the roots is symbiotic. This kind of symbiosis boosts the soil ecology and has its bearing on primary productivity.

In ectomycorrhizas (sheathing mycorrhizas), the symbiotic association is established between a Basidiomycete fungus and a tree. In this, a mantle of the fungus is formed around the root that serves as a network involving intercellular hyphae in the cortex and a ramifying mycelium in the soil. There are no intracellular connections in this kind of root–fungus association, and the roots are normally stunted.

Endomycorrhizas are of three types: the first two types found in the order Ericales and the family Orchidaceae, respectively, and are very specialized in their nature and functions, while the third group known as vesicular-arbuscular (VA) mycorrhiza is formed with Phycomycetous fungi.

Most families of the higher plants, Pteridophytes and liverworts in the Bryophytes, are associated with VA mycorrhizas. This relationship is intracellular with fungus-producing haustoria that invaginate the plasmalemma and produce what is known as arbuscules, hyphae, and vesicles.

Mycorrhizas are capable of enhancing P uptake and, thus, improving plant growth. Mycorrhizal plants grow much better in the P-deficient soils than those that do not have mycorrhizal association. However, in the P-rich soils, the case is reverse. Mycorrhizal roots are normally shorter than the non-mycorrhizal ones. It is because compared to the non-mycorrhizal roots, the roots with mycorrhizal association absorb P more quickly per unit root length. The fungus can utilize the P source that the plant cannot and travels to the root via the fungus. The mycorrhizal root, in this way, has a different kinetics for P uptake. VA mycorrhizas explore soil and harness nutrients, which do not become accessible to plants.

SUMMARY

The pedosphere is composed of mineral, fluid, gaseous, and biological components and is a repository of most of the macro and micronutrients essential for plants. The sedimentary nutrient cycles operate in the soil. Gaseous cycles too have strong connections with the pedosphere. Soil is an ecosystem, far more complex than other ecosystems on Earth. The biodiversity of the soil ecosystems is richer than that of the above-ground ecosystems. The diversity of soil (pedodiversity) is the manifestation of the diverse natural features of the earth. Soils of the world vary according to their geographical positioning/climatic region, physical structures, chemical composition, and biological features. Climatic or environmental factors, seasons of the year, geologic factors, and hydrologic factors are also the dominant factors that determine the physicochemical and biological properties of the soils. There are six main soil types planet Earth exhibits: sandy, silty, clay, loamy, peaty, and chalky. These soil types are made up of three major soil types, viz., sandy, silty, and clay. Soils host a quarter of our planet's biodiversity. Pedodiversity (variety of soils) and biodiversity (variety of living organisms) strongly correlate with each other at the global level.

There are some 16 elements regarded as essential for all the plants we cultivate: C, H, O, N, P, K, Ca, Mg, S, Cu, Zn, Mn, Fe, B, Cl, Mo. However, Na, Si, and Co are also essential for some plants. Among these nutrients, some needed by plants in relatively large quantities (C, H, O, N, P, K, Ca, Mg, S, and Fe), are called major elements, or macronutrients. Some elements, on the other hand, are required by the plants in relatively smaller quantities (Zn, Cu, Mn, B, and Mo) and are called microelements, or micronutrients or trace elements.

Plant-nutrient sources include inorganic and organic substances. These nutrients come from soil, water, and atmosphere—the components of the biosphere. Nutrients in water (as soil solution) are present in the form of cations and anions. Plants depend only on inorganic nutrients, making a "bridge" between the inorganic environment and organic world.

Generally, soil pH between 6.0 and 7.5 is appropriate for making most of the nutrients available for the plants. Alkaline soils are appropriate for increasing Ca, Mg, Mo, and K availability for the plants. Such soils reduce the availability of Zn, Mn, and B. Acidic soils increase the solubility of Mn, Al, and Fe, which, in excess, are toxic for the plants. A common chelating agent used for Zn, Fe, and other micronutrients is EDTA. The SOM helps overcome or counteract certain toxic substances (or soil pollutants).

All nutrients in a soil solution do not function independently and do not become available to plants in proportion of their individual quantities. Availability of an ion is influenced—enhanced or suppressed—by the presence of other ion(s). Deficiency of some nutrients might also occur due to mixing up and bonding between two (or more) elements, which make a salt that cannot be absorbed by plant roots. The sum of equivalents of cations in a soil solution is equal to the sum of soluble anions.

Soil constantly supplies ions to the plants due to which vegetation flourishes in terrestrial ecosystems and a variety of communities comes into being. However, mineral nutrients do not occur in the soils in a fixed amount. They are dynamically added to or are withdrawn from the soil but tend to maintain an equilibrium, which is influenced by a number of environmental factors, both biotic and abiotic. Apart from N, many other minerals also recycle through SOM. A variety of minerals are there in the freshly fallen leaves, which, through decomposition, are added to the soil nutrient pool.

The root cells of a plant contain a much higher concentration of ions than the surrounding medium. In comparison to their surrounding soil solution, the root cells are also electrically negative. Most cations in the surrounding medium, as a consequence, enter the cells passively. The passive transport of an ion is down the electrochemical gradient but usually against the chemical concentration gradient for that ion. Cations, in exchange of protons in some of the cases, especially for Na^+, actively export back to the soil solution. The intercellular transport of ions in plant roots takes place through the plasmodesmata, the living connectors between the cells, and this is called the symplast pathway. Occurring through a concentration gradient, the ions passively diffuse through the cells of the cortex and endodermis to be eventually delivered into the xylem vessel or tracheids. The movement of the ions from cell to cell across their plasma membrane actively and passively through the cortex and epidermis cells is called the transmembrane pathway. Furthermore, the movement of ions through cell walls and intercellular spaces between the cells in the cortex is called the apoplast pathway.

The rate of root uptake of the ions increases with an increase in ion concentration. But with continuous increase when the kinetics saturation is reached, the ion uptake becomes independent of ion concentration. Ion transporters are the membrane-bound proteins with a high degree of ion specificity. The cation-exchange capacity of the root-influencing nutrient uptake varies with plant species, variety (genotype) of the species, and age of the plants. The vacuole of the cell plays a crucial role in striking a balance for stabilizing the cellular demand and supply of ions.

A number of environmental factors and internal factors influence ion transports, both passive and active, between the soil solution and plant roots. Nutrient uptake by plants is influenced by the root:shoot ratio and root:weight ratio, root diameter and root hairs, root density and distribution, rhizosphere, and microbial populations in the soil. N fixation in the soil is influenced by environmental factors such as CO_2, C:N ratios, weather factors (heat, drought, soil moisture), soil pH, mineral nutrients, and mycorrhizas.

REFERENCES

Aboulroos, S. A. and Nielsen, N. E. 1979. Mean rate of nutrient uptake per unit of root and finess, length and density of barley roots in soil at various phosphorus levels. *Acta Agric. Scand.* 29: 326–330.

Bananomi, G., Cesarano, G., Gaglione, S. A., Ippolito, F., Sarker, T. C. and Rao, M. A. 2016. Soil fertility promotes decomposition rate of nutrient poor, but not nutrient rich litter through nitrogen transfer. *Plant Soil.* 412: 397–411. doi:10.1007/s11104-016-3072-1.

BassiriRad, H. 2000. Kinetics of nutrient uptake by roots: Responses to global change. *New Phytol.* 147: 155–169.

Becking, J. H. 1956. On the mechanism of ammonium uptake by maize roots. *Acta Bot. Neerl.* 5: 2–79.

Behl, R., Tischner, R. and Raschke, K. 1988. Induction of a high capacity nitrate uptake mechanism in barley roots prompted by nitrate uptake through a constitutive low-capacity mechanism. *Planta* 176: 235–240.

Bloom, A. J., Chapin, F. S. and Mooney, H. A. 1985. Resource limitation in plants–an economic analogy. *Ann. Rev. Ecol. Syst.* 16: 363–392.

Brownell, P. F. 1965. Sodium as an essential micronutrient element for a higher plant (*Atriplex vesicaria*). *Plant Physiol.* 40: 460–468.

Brownell, P. F. and Crossland, C. J. 1972. The requirement for sodium as a micronutrient by species having the C_4 dicarboxylic photosynthetic pathway. *Plant Physiol.* 49: 794–797.

Claassen, N. and Barber, S. A. 1977. Potassium influx characteristics of corn roots and interaction with N, P, Ca and Mg influx. *Agronomy J.* 69: 860–864.

Clark, F. F. 1949. Soil microorganisms and plant growth. *Adv. Agron.* 1: 241–288.

Doddema, H. and Telkamp, G. R. 1979. Uptake of nitrate by mutants of *Arabidopsis thaliana*, disturbed in uptake or reduction of nitrate. II. *Kinetic. Physiol. Plant.* 45: 332–338.

Drew, M. C., Saker, L. R., Barber, S. A. and Jenkins, W. 1984. Changes in the kinetics of phosphate and potassium absorption in nutrient-deficient barley roots measured by a solution-depletion technique. *Planta* 160: 490–499.

Elawad, S. H., Gascho, G. J. and Street, J. J. 1982. Response of sugarcane to silicate source and rate. *Agronomy J.* 74: 481–484.

Epstein, E. 1976. Kinetics of ion transport and the carrier concept. In: Luttge, U., Pitman, M. G. (eds.) *Encyclopedia of Plant Physiology, New Series II B.* Berlin, Germany: Springer-Verlag, pp. 70–94.

Fageria, N. K., Baligar, V. C. and Jones, C. A. 2011. *Growth and Mineral Nutrition of Field Crops*, 3rd edn. Boca Raton, FL: CRC Press, Taylor & Francis Group.

FAO. 2015. *Soils and Biodiversity: Soils Host a Quarter of Our Planet's Biodiversity–2015 International Year of Soils.* Rome, Italy: FAO.

Fitter, A. H. and Hay, R. K. M. 2002. *Environmental Physiology of Plants*, 3rd edn. London, UK: Academic Press. 367 p.

Gardner, F. P., Pearce, R. B. and Mitchell, R. L. 2018. *Physiology of Crop Plants*. Jodhpur, India: Scientific Publishers. 327 p.

Giller, K. E. and Day, J. M. 1985. Nitrogen fixation in the rhizosphere: Significance in natural and agricultural systems. In: Fitter, A. H., Atkinson, D., Read, D. J. and Usher, M. B. (eds.) *Ecological Interactions in Soil*. Oxford: Blackwell Scientific Publications, pp. 127–147.

Gutschick, V. P. and Kay, L. E. 1995. Nutrient-limited growth rates: Quantitative benefits of stress responses and some aspects of regulation. *J. Exp. Bot.* 46: 995–1009.

Halina, R. E., Brown, L. K., Benaouah, A. G., Youna, I. M., Hallet, P. D., White, P. J. and George, T. 2013. Root hairs improve root penetration, root-soil contact, and phosphorus acquisition in soils of different strength. *J. Exp. Bot.* 64 (12): 3711–3721.

Ibáñez, J. J. and Feoli, E. 2013. Global relationships of pedodiversity and biodiversity. *Vedose Zone J.* 12 (3): 10–21.

Inomura, K., Bragg, J. and Follows, M. J. 2017. A quantitative analysis of the direct and indirect costs of nitrogen fixation: A model based on *Azotobacter vinelandii*. *ISME J.* 11 (1): 166–175.

IPCC. 2013. The Physical Science Basis–Technical Summary (PDF). Intergovernmental Panel on Climate Change. 89–90.

Jakobsen, S. T. 2009. Interaction between plant nutrients: 1. Theory and analytical procedures. *Acta Agr. Scand. B-S. P.* 42 (4): 208–212.

Johnson, I. R. 1985. A model of the partitioning of growth between the shoots and the roots of vegetative plants. *Ann. Bot.* 55: 421–431.

Katznelson, H. 1946. The 'rhizosphere effect' of mangels on certain groups of soil microorganisms. *Soil Sci.* 62: 343–354.

Pilbeam, D. J. and Kirsby, E. A. 1990. The physiology of nitrate uptake. In: Abrol, Y. P. (ed.) *Nitrogen in Higher Plants*. New York: John Wiley & Sons, pp. 39–64.

Rietra, R. P. J. J., Heinen, M., Dimkpa, O. and Bindraban, P. S. 2017. Effect of nutrient antagonism and synergism on yield and fertilizer use efficiency. *Commun. Soil Sci. Plan.* 48 (16): 1895–1920.

Risser, P. G. 1985. Toward a holistic management perspective. *BioScience* 35: 414–418.

Sainju, U. M., Allen, B. L., Lenssen, A. W. and Ghimire, R. P. 2017. Root biomass, root/shoot ratio, and soil water content under perennial grasses with different nitrogen rates. *Field Crop Res.* 210: 183–191.

Schatz, A., Schalscha, B. and Schatz, V. 1964. Soil organic matter as a natural chelating material: The occurrence and importance of paradoxical concentration effects in biological systems. *Compost Sci.* 5 (1): 26–30.

Singh, V. 2016. An amazing world beneath your feet. *The Speaking Tree*. March 19, 2014.

Sohm, J. A., Webb, E. A. and Capone, D. G. 2011. Emerging patterns of marine nitrogen fixation. *Nat. Rev. Microbiol.* 9: 499–508.

Starkey, R. L. 1929. Some influences of the development of higher plants upon the microorganisms in the soil. *Soil Sci.* 27: 319–334.

Sukumaran, D., Anilkumar, A. and Thanga, S. G. 2015. Root soil (R/S) ratio in plants used for phytoremediation of different industrial effluents. *Int. J. Environ. Bioremediat. Biodegrad.* 3 (1): 10–14.

Tingey, D. T., Phillips, D. L. and Johnson, M. G. 2000. Elevated CO_2 and conifer roots: Effects on growth, life span and turnover. *New Phytol.* 147: 87–103.

Treseder, K. K. and Allen, M. F. 2000. Mycorrhizal fungi have a potential role in soil carbon storage under elevated CO_2 and fluctuating nitrogen deposition. *New Phytol.* 147: 189–200.

Van Cleve, K., Barney, R. and Schlentner, R. 1981. Evidence of temperature control of production and nutrient cycling in two interior Alaska black spruce ecosystems. *Can. J. For. Res.* 11: 235–273.

Van den Honert, T. H. and Hooymans, J. M. 1955. On the absorption of nitrate by maize in water culture. *Acta Bot. Neerl.* 4: 376–384.

Williams, M. C. 1960. Effect of sodium and potassium salts on growth and oxalate content of Halogeton. *Plant Physiol.* 35: 500–505.

WEBSITES

http://doi.org/10.1080/00103624.2017.1407429

http://nutriag.com/article/mulderschart

http://theconversation.com/healthy-soil-is-the-real-key-to-feeding-the-world-75364?utm_source=linkedin&utm_medium=linkedinbutton

http://www.fao.org/3/a-i4551e.pdf

http://www.lawsonfairbank.co.uk/soil-types.asp
https://dl.sciencesocieties.org/publications/vzj/abstracts/12/3/vzj2012.0186
https://doi.org/10.1016/j.fcr.2017.05.029
https://doi.org/10.1080/09064719209410213
https://doi.org/10.1093/jxb/ert200
https://www.maximumyield.com/how-nutrient-antagonism-leads-to-nutrient-deficiency-in-plants/2/2092
https://www.ncbi.nlm.nih.gov/pmc/articles/PMC5315487/
https://www.ncbi.nlm.nih.gov/pmc/articles/PMC5610682/
https://www.soils.org/discover-soils/story/diversity-soil-types-can-be-indicator-aboveground-biodiversity
https://www.speakingtree.in/article/an-amazing-world-beneath-your-feet
https://www.thedailygarden.us/garden-word-of-the-day/mulders-chart

4 Water Relations

Water is the driving force of all nature.

Leonardo da Vinci

THE WATER PLANET AS HOME TO LIFE

Our Earth is a water planet. Water is not just a major component of the environment on Earth, but also a major component of all living organisms. The proportion of water on Earth and in living organisms is almost equal. This fact reveals that the phenomenon of life is uniform all over the planet as well as across all living organisms. Water is both an ingredient in anabolism as well as a product of catabolism. Water is central to the functioning of life because all the enzymes function only in an aquatic medium. Also, the role of water is vital in maintaining a thermal budget of the biosphere. The polarity of water makes it an excellent solvent.

Despite having the largest share in the constitution of the living planet, the distribution of water on Earth is unequivocally uneven in terms of space, time, and type. Therefore, availability, acquisition, and conservation of water are important considerations, especially in the context of the organisms in desert, air, and semiarid environments.

All organisms conserve an optimum amount of water in their body structures, so that all life processes continue to operate, keeping the organisms active and functional. Water loss from the body is compensated with water intake by organisms. Organisms also live in aquatic environments where there could be a danger of water flooding into their bodies. But it does not take place, and the aquatic organisms maintain their water balance. Yet there are organisms that live in environments with high salinity, and they are successful in preventing saline water from flooding their bodies. Most of the organisms would expend energy in order to maintain their water balance in their environment. How the organisms maintain water balance is a subject of their water relations. An understanding of water relations is interesting to understand how the organisms are adapted to different environments characterized by water scarcity, water excess, and water type as well as how the water mediates life processes among organisms.

WATER PROPERTIES CONDUCIVE TO LIFE

From the Himalayan and polar ice caps to the steamy geysers, water is found almost everywhere on planet Earth. And where there is water, there is life. Water is such a wonderful ingredient and medium of life! Follow water and you will follow life. Microorganisms would be found perpetuating invariably everywhere wherever there is water—ranging from the ice-capped polar lakes to hydrothermal vents on the ocean floors. These microorganisms are the primary symptoms of the intricate water–life association. What are the qualities, properties, and characteristics of water—the "elixir of life"—that are indispensable for life, vital to life?

Water is the only compound that exists simultaneously in the three states of the matter: solid, liquid, and gas. The water molecule is outstanding in its structure. Two hydrogen atoms are bonded to an oxygen atom, and this bonding is such that the water becomes a universal solvent. The excellent solvent quality of water is because of the high degree of polarity a water molecule possesses. The two hydrogen atoms bunch on one side of the molecule, contributing to create a positive region. The oxygen end of the molecule, on the other hand, has a negative charge. The hydrogen atom ends tend to attract all the atoms with an extra electron in its outer shell (or the negative ions). The oxygen atom end of the water molecule attracts the atoms with one of their stripped-off electrons (or the positive ions). With its astounding dissolving properties, water serves as a perfect medium for the movement of substances from one place to another. It is thanks to this property that water plays a crucial role in the biogeochemical cycles and in transmitting nutrients (such as Na^+, Cl^-, Ca^{2+}, and PO_4^- ions) into and out of a cell.

The capillary action of water is a unique property that even defies gravity. This property makes water maintain its reserves of soil vital for the life of the plants. Again, this property also helps water movement through the xylem of plants. The capillary water action, thus, ensures continuity of photosynthesis the water is an indispensable input of.

The unusual density of ice is yet another characteristic of water. As water cools to the freezing point, water molecules crystallize and expand. This expansion of water upon its solidification is known as anomalous expansion. The density of the solid water (ice) is lower than that of the water, making it float on liquid water. This property of water protects aquatic life in polar and temperate regions of the earth. The solid ice in the winter season in the temperate regions makes a layer on natural lakes and other water bodies. A layer of the ice over a water body insulates the liquid water below, not allowing the temperature to fall below a point that the inhabited organisms won't survive in. Due to this process, aquatic organisms thrive comfortably in the water below the ice cover even when the atmospheric temperatures are extremely low. Aquatic animals in the polar region survive due to this property of water.

The high-heat capacity of water helps living organisms regulate their body temperatures. Water needs to absorb very high amounts of energy to increase temperature and transform its liquid state into vapor. Water in the cells of all the living organisms, thus, acts as a buffer to resist change in temperature and maintain homeostasis. In other words, water helps the organisms acclimate and adapt to live amidst the prevailing conditions of changing environmental temperatures.

WATER AVAILABILITY TO ORGANISMS

Water availability for organisms in an environment depends on a number of factors. The first and the foremost factor is the way water movement takes place. Following the universal law, water moves down concentration gradients. Whether the organism loses or gains water depends on the water-concentration gradients between organisms and the environment. One of the indicators as well as determinants of water availability in the environment is the climate or microclimate of a region or an area. The organisms in tropical rainforests with high humidity experience excessive amounts of water available for them. The organisms in the arid environments, on the other hand, would face acute water scarcity. The water content of air, or relative humidity, is one of the basic characteristics determining movement of water due to change in the magnitude of the gradients and in imparting microclimate to an area.

WATER IN AIR

On the water planet, water evaporates continuously from oceans, seas, lakes, and rivers at all temperatures and enters into the atmosphere. A significant proportion of the evaporated water is contributed by living organisms. The loss of water from the water sources/organisms to the atmosphere depends on the water-concentration gradient between the two. As the water content in the atmosphere goes on increasing, the water gradient between the hydrosphere/living organisms and the atmosphere goes on decreasing, leading to decreased water loss to the atmosphere. In humid climate, water loss to the atmosphere is much less than that in the arid climate.

Thus, it is water quantity (water vapor) in the atmosphere that matters. Quantity of the water in the atmosphere is conveniently expressed in a relative term. Air seldom contains all the water it can hold; we can use its degree of saturation with water vapor as a relative measure of water (vapor) content. The most commonly used measure is the relative humidity, that is, the water content of air relative to its content at saturation, expressed as follows:

$$\text{Relative humidity} = \frac{\text{Water vapor density}}{\text{Saturation water vapor density}} \times 100$$

The water-vapor density—mass of water vapor per unit volume of air—is measured in H_2O (mg)/L (milligram of water per liter of air) or H_2O (g)/m^3 (gram of water per cubic meter of air). The saturation water-vapor density, which is the amount of water that the air can potentially hold, changes with the air temperature. Hot air has more potential to hold water vapor than cold air.

The water content in the air can also be expressed in terms of the pressure it exerts. The total atmospheric pressure involves all the gases in the air. But we take into consideration only the water vapor; we call it water-vapor pressure. The pressure the water exerts on air saturated with water is the saturation water-vapor pressure. The amount of the pressure is directly proportional to its density and increases with an increase in temperature.

At sea level, the atmospheric pressure (measure of all the gases in air) averages approximately 760 mm Hg. The international convention to represent water-vapor pressure is pascal (Pa).

1 Pa = 1 Newton (1N) of force applied over an area of 1 m^2, that is, 1 Pa = 1Nm^{-2}
1 atmospheric pressure = 760 mm Hg = 101,300 Pa = 101.3 kPa = 0.101 MPa

Atmosphere is not always saturated with water. When it becomes saturated, water will condense out in clouds, dew, or film of water on leaves. Otherwise, there is a difference between the amount of water the air can potentially hold and the amount of water at a particular point of time. This difference is known as the vapor-pressure deficit (VPD). Water flow from an organism to the atmosphere in terrestrial environments takes place in accordance with the VPD of the air surrounding the organism.

WATER MOVEMENT WITHIN WATER

Asking about how water movement takes place within water (i.e., in aquatic environments) appears to be a joke, but it is true that water movement has a definite direction conditioned by the concentration of water. Concentration of water? Is it a true notion? Yes, the concentration of water is a decisive factor determining the direction of water flow in aquatic environments.

We know that pure water without any dissolved salts in it is very rare in nature. As the purity of water is reduced due to dissolved salts, its concentration also decreases. Thus, water in a freshwater source is more concentrated than in the oceans. The latter have more dissolved salts than the former; hence, water gets diluted, or in other words, there is less water per unit volume in marine ecosystems than in the freshwater sources. Further, water in the Great Salt Lake or in the Dead Sea is more dilute than in the oceans. It is due to the dissolved substances that dilution of water increases. The greater the quantities of dissolved solutes in water the higher the degree of water dilution.

What happens in the case of the organisms living in water? Their body fluids containing water too have dissolved inorganic as well as organic substances. Hence, the concentration of the fluids within the body of the organisms and that of their external environment, in most of the cases, is likely to be different. If the water flows along its concentration gradient, as it has to happen, the aquatic organisms would either be flooded with water, or their body fluids will move to their external aquatic environment. Under such circumstances the aquatic organisms cannot survive. But since the two environments are separated by a selectively permeable membrane, the movement of water between the two environments is controlled by what is known as osmosis—the diffusion of water across a semipermeable membrane.

Depending on the amount of water per unit volume (or, in other words, concentration of the water), the aquatic organisms can be divided in three categories:

1. Isosmotic: Concentration of water in the body fluids and external environment is alike; there is no movement between organisms and their environment.
2. Hypoosmotic: Concentration of water in body fluids is higher than in the external environment; organisms tend to lose water to the external environment.
3. Hyperosmotic: Concentration of water in body fluids is lower than the external medium; organisms tend to be water-flooded due to movement of external water toward their body fluids.

When moving along its concentration gradient, water produces osmotic pressure, which, like vapor pressure, is also measured in pascals (Pa). The difference in the water concentration across a semipermeable membrane determines the strength of the osmotic pressure. The larger the differences in the concentration of the fluids between the two aquatic environments across a semipermeable membrane the higher the osmotic pressure generated. In order to maintain a proper internal environment, the aquatic organisms must expend energy. The amount of the energy expended by the aquatic organisms is proportional to the magnitude of the osmotic pressure between them and their environments.

WATER POTENTIAL

In terrestrial ecosystems, the VPD of the air surrounding an organism influences water movement from the organism to the atmosphere. In aquatic ecosystems, however, water moves both sides: either from an organism to a water body, or from the water body to an organism subject to water or solute concentrations of the organism's body fluids and that of the medium of the water body. Of course, the water movement between the organisms and the water bodies is in accordance with concentration gradient—that is, the water movement down its concentration gradient: from hypoosmotic to hyperosmotic direction.

Water movement within the soils takes place through the small pore spaces and within plants through the small water-conducting cells. Water movement from the soil through a plant and into the atmosphere takes place down the gradient of water potential.

Water potential is the capacity of water to do work. Flowing water has its kinetics to do work. For example, it can propel the wheel and help operate a grinding machine. It can also be instrumental in turning the turbines and generating electricity. Water flows, as we experience in mountain areas, from positions of higher free energy to lower free energy. The capacity of water to do work also depends on its free energy content. The flow of water from the hypoosmotic to the hyperosmotic direction, that is, down osmotic gradients, reveals that it has capacity to do work. The hypoosmotic water has more free energy than the hyperosmotic water. The hypoosmotic water is purer than the hyperosmotic water. Thus, in other words, we can say that the pure water has more free-energy content than the sea water. The movement thus takes place from pure water to saline water. This water movement generates osmotic pressure, which is measurable.

Water potential—that is, the potential energy of water—is measured in pascals, usually megapascals:

$$MPa = Pa \times 10^6$$

Water potential is denoted by the symbol ψ (psi). The water potential of pure water is set at 0, which is the highest value of water potential in nature.

Pure water with 0 Pa is very rare in nature. Because water is a universal solvent, it dissolves some solutes almost everywhere. Even rainwater is not pure in itself as it would also dissolve some substances in it. Thus, water potentials in nature are by and large negative. Thus, water movement takes place down a water-potential gradient: from less negative to more negative. For example, water in the soil is higher, that is, less negative; in the plant body, it is more negative; and in the dry air, it is of the highest negative value. Thus, water moves from soil, through the plant, to the air, that is, from a less negative to a more negative direction of water potential.

The gradient of water potential is produced by certain mechanisms. The water potential of a solution is expressed as:

$$\Psi = \Psi_{solutes}$$

$\Psi_{solutes}$ denotes the reduction in water potential due to dissolved solutes. $\Psi_{solutes}$ has a negative value.

Solutes are not the only factor reducing water potential. Within the pore spaces and small spaces in the interior of plant cell, matric forces are also at work. As water tends to adhere to the walls of a container, as in the case of the cell walls or the soil particles lining a soil pore, it lowers the value of water potential due to matric forces. Thus, the water potential of the fluids within plant cells is approximately:

$$\Psi_{plant} = \psi_{solutes} + \psi_{matric}$$

In the preceding equation, ψ_{matric} denotes the reduction in water potential due to matric forces within plant cells. As the evaporation of water from the leaf surface takes place, it generates yet another force: a pressure, or tension, on the water column in the whole plant from the leaf surface, through the trunk, down to the roots. The water potential is further reduced due to this negative pressure. Water potential, thus, is affected (reduced) by solutes, matric forces, and negative pressure:

$$\Psi_{plant} = \psi_{solutes} + \psi_{matric} + \psi_{pressure}$$

$\psi_{pressure}$ is the reduction in water potential due to negative pressure exerted by water evaporation from the leaf surface.

The water movement behavior in soil is somewhat complex. The concentration of solutes in soil water is very low, allowing the matric forces to play a major role in determining water potential.

The matric forces vary according to the soil type, mainly soil texture and pore size. With larger pore size—sand and loam soil, for example—the matric forces will be less. And if the pore size is smaller—as in the case of clay soil—the matric forces will be of higher magnitude. It is why the coarser soils with lower matric forces do not hold water as tightly as the clay soil that holds more water for a longer period. The smaller pore size and, therefore, high magnitude of matric forces exerted by clay soil, causes some resistance in water movement within the soil. The water flow from the soil to a plant will continue as long as water potential of the soil is higher than that of the plants, that is, $\Psi_{soil} > \psi_{plant}$.

As water moves from the soil to the roots due to $\Psi_{soil} > \psi_{plant}$, it joins a column of water extending from the roots through the xylem (water-conducting cells) to all parts of the shoot of a plant. In the water column, the water molecules are bound with each other with the help of hydrogen bonds. The water molecules at the surface of the leaves evaporate into the atmosphere, which, in turn, exerts a negative pressure, or tension, on the water column in the entire plant. The continuous evaporation of water dilutes water (i.e., increases solute concentration) in plant fluids leading to a further decrease in ψ_{plant}, which contributes to creating more steepness in the water-potential gradient between the soil and plant (see Figure 4.1).

In a dry season when water evaporation from the leaf surface is likely to increase, plants meet up their water demand by extracting more water from the soil, and in this process, a steeper water-potential gradient from the soil to the plants is of great help. However, in a dry season, if water is not supplied continuously, there is more probability of the soils becoming water deficient due to the extraction of more amounts of water from the soil by a plant coupled with increased water evaporation from soil surface. The water is first extracted from the larger soil pore spaces due to

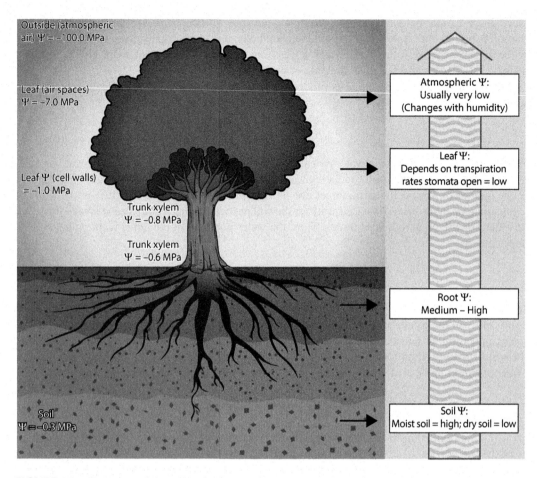

Outside (atmospheric
air) $\Psi = -100.0$ MPa

Leaf (air spaces)
$\Psi = -7.0$ MPa

Leaf Ψ (cell walls)
$= -1.0$ MPa

Trunk xylem
$\Psi = -0.8$ MPa

Trunk xylem
$\Psi = -0.6$ MPa

Soil
$\Psi = -0.3$ MPa

Atmospheric Ψ:
Usually very low
(Changes with humidity)

Leaf Ψ:
Depends on transpiration
rates stomata open = low

Root Ψ:
Medium – High

Soil Ψ:
Moist soil = high; dry soil = low

FIGURE 4.1 Water potential gradient from the soil through the roots, trunk, and leaves to the atmospheric air (values are approximate).

lower matric forces (and more water potential), leaving water in the smaller pore spaces. As water concentration in soils decreases (and solute concentration increases) considerably, water potential in the soils drops (becomes more negative), making it harder for the plants to draw water from the soils.

PLANT ROOTS AND WATER ACQUISITION

There is a definite relationship between root development and water acquisition by the plants. Water availability in an ecosystem greatly influences root development. Reciprocally, the extent of root growth greatly influences water accessibility to the plants. The extent of root development in plants reflects differences in the availability of water in habitats. If water availability in a habitat is plentiful, root development in native plants might not be as much as in case of their counterparts in the arid and semiarid areas.

Root development is also associated with climate. The climate itself determines the availability (or deficiency or unavailability) of water for the plants. Plants occurring in dry climates tend to develop deeper root systems than those in moist climates. The root biomass proportion of a plant, thus, is also greater in plants occurring in dry climates than those in moist climates. Wherever the water table is very deep, as in the deserts, tap roots of some shrubby species can extend up to 30 m down into the soil. Extra growth of the roots in water-scarce habitats helps the plants provide them access to water available in deeper layers of the land. Coniferous forests harbor trees with about 25%

TABLE 4.1

Effect of Different Levels of Drought Stress on Root:Shoot Ratio by Length and Biomass

	Root:Shoot Ratios by Length		Root:Shoot Ratios by Biomass	
Drought Levels	Black Finger Millet	Black Finger Millet	Black Finger Millet	Black Finger Millet
Control	0.233	0.241	0.194	0.202
15 days	0.331	0.347	0.172	0.194
30 days	0.737	0.753	0.162	0.178
45 days	1.923	1.957	0.105	0.135

Source: Khatoon, H. and Singh, V., *Environ. Ecol.*, 35, 1595–1604, 2017.

root biomass of total plant biomass. In extremely dry environments with very low water tables and in arid and semiarid grasslands, roots may grow to the extent that as high as 90% of the total plant biomass would be accumulated in the roots.

An experiment conducted by Khatoon and Singh (2017) to observe the effects of water stress on the root:shoot ratio in two races of finger millet (*Eleusine coracana* L.) unfolded the fact that the root:shoot ratio by length increases with an increase in drought spell. However, the root:shoot ratio by weight decreases with furthering of drought period (Table 4.1). It can be elucidated from this experiment that water scarcity has a negative effect on the overall growth as gauged by biomass accumulation, but the length of the roots increases to maintain access to water availability to plants. An increase in root length in search of water from deeper soil layers fails to compensate for the normal growth of the plants.

Only those plants that can grow their roots deeper enough into soil profile to harp on soil moisture can survive in drier habitats. The greater root biomass contributes to maintain stable and higher leaf water potential in the plants growing in dry habitats. The lower root biomass, on the other hand, tends to decrease leaf water potential (Molles 2005).

Soil moisture influences root growth of the plants. Findings of Schenk and Jackson (2002) based on a study of a set of 475 root profiles from 209 places across the globe revealed that water-deficient ecosystems caused the inhabited plants to grow their roots deeper into the soil profile. Their analysis of a massive set of plants from across the earth also brought to the fore the fact that an increase in rooting depth takes place from 80° to 30° latitude, that is, from the Arctic Tundra to the Mediterranean woodlands/shrublands and deserts.

WATER RELATIONS OF PLANT CELLS

The movement of water into and out of plant cells is determined by certain factors, biotic as well as abiotic. An understanding of these environmental factors is crucial because they influence turgidity and growth. Root pressure, guttation, and exudation from wounds in the plants are also phenomenally influenced by water relations of plant cells. The osmotic concept proposed in the first half of the nineteenth century was considered pertinent to the explanation of the water relations of plants. Proposed by Dutrochet in 1837, this classical concept of cell–water relations regarded the cell as an osmometer allowing the protoplasm to function as a "passive, differentially permeable membrane separating the vacuole sap from external solution" (Kramer and Currier 1950). According to this concept, water movement was regarded to occur solely along osmotic gradients without any activity of protoplasmic membranes. Later on, the modified concept stated that cell-to-cell water movement takes place along diffusion-pressure gradients, not along osmotic pressure or concentration gradients.

The plant cells that play roles in maintaining plant–water relations are mature and have greater fractions of cell water in their central vacuoles. The vacuolar contents of a plant cell are separated from the external environment by means of a thin layer of cytoplasm along with the plasmalemma and tonoplast that function like a complex semipermeable membrane.

The external environment of a leaf includes cell walls and intercellular spaces of the leaf, or leaf apoplasts, which are subject to atmospheric pressure, that is, $\Psi_p = 0$. The solute concentration is quite low, contributing to a lower value of Ψ_s. Matric forces exerted by the cell walls, therefore, determine water potential in the apoplast (Ψ_{apo}). Thus,

$$\Psi_{apo} = \Psi_m$$

Well-supplied with water and not transpiring (at night), a leaf will normally have a high Ψ_m value (−0.1 MPa). Matric forces, however, are hardly of much consideration in the case of the cell vacuole. Because a solute is essentially present in a vacuole, here water potential is determined mainly by the solute concentration. Therefore,

$$\Psi_{vac} = \Psi_s$$

Here Ψ_s can range between −0.5 and −3.0 MPa. In some cases, it goes even lower than this, and that would depend on the species and osmoregulation.

Since $\Psi_{vac} < \Psi_{apo}$, water tends to flow inward across the cytoplasm, contributing to an increase in water concentration (decreasing solute concentration) of vacuolar sap. Water flow toward the vacuole also results in an increased vacuolar volume. If a plant cell is lacking a cell wall and if the water potential difference is not abolished, the continuous flow of water along the water-potential gradient (i.e., from apoplast to cell vacuole) will result in the bursting of the cell vacuole. In a leaf cell, the volume of the vacuole is limited by its cell walls. Only limited water flow is maintained thanks to the cell wall elasticity. As a consequence, the vacuole builds up turgor pressure (hydrostatic pressure) that exerts pressure on the inner surface of the cell walls, resulting in an increase of the vacuolar water potential. The increased turgor pressure of a vacuole puts pressure on the adjacent cells. Thus, the whole leaf attains a state of turgidity. When water influx (difference in solute potential) and water efflux (cell turgor pressure) are equal, an equilibrium is reached with no further water movement into the vacuole. Cell turgor pressure at this juncture is maximum, and then,

$$\Psi_{apo} = \Psi_{vac}$$

or

$$\left(\Psi_m\right)_{apo} = \left(\Psi_s + \Psi_p\right)_{vac}$$

As almost all the physiological and biochemical processes take place in the cytoplasm or in cell organelles, water relations of cytoplasm are of critical interest. Because of the thermodynamic equilibrium within a cell, there is much likelihood that water potential in the cytoplasm (Ψ_{cyt}) equals to that in a vacuole (Ψ_{vac}), that is,

$$\Psi_{cell} = \Psi_{cyt} = \Psi_{vac}$$

SUPPLY OF WATER BY THE SOIL

Soil water is absorbed by the plants through the root system. Water absorption through leaves and/ or other parts of the shoot system is ruled out in most of the plants in terrestrial systems. However, some plants in arid areas also absorb rainwater drops or dew by leaves, which is critical for their survival in the extremely inhospitable climates.

Soil holds certain amounts of water. The quantity of water in a particular soil depends on climate, and precipitation in excess of evapotranspiration (i.e., P–E). In areas where P > E, water availability for plants is no problem. The greater the difference between P and E, the greater the availability of soil water for the plants. Such "extremely humid" areas, for example, the rainforests, are quite appropriate for vegetation growth and plant distribution. In sharp contrast are the areas where the annual precipitation is less than the soil water lost through evapotranspiration (P < E). In these "extremely arid" areas, such as the Thar Desert of India, water paucity for the plants is usually critical for their survival. Vegetation in the P < E-ridden areas is less dense, and plant distribution is sparse. Vast patches of the land in extremely dry areas might be devoid of any vegetation.

Apart from the extreme habitats are the humid and arid habitats in which availability of soil water, and therefore performance of the vegetation, will depend on P–E as well as on annual distribution of rainfall.

What are the major forces of retaining water in soil pores? As the soil-solute concentration is generally low, matric forces happen to be the major force to retain water in the soil pores. There is an inverse relationship between the matric force and pore diameter (d) size. As the pore size decreases, matric forces increase. The water potential of the water in the soil pores is inversely related to the pore diameter. The pore water potential (Pa) and pore diameter (d in μm) relationship is expressed as follows:

$$\Psi_{\text{pore water}} = \Psi_m = -0.3/d$$

For instance, pure water held in soil pores of diameter 30 μm will be at a potential of 0.01 MPa. In order to withdraw water from the pores of 30 μm, it will require a suction of at least 0.01 MPa. Webster and Beckett (1972) found that in temperate regions, gravity exerts a suction of 5 kPa; all soil pores wider than 60 μm will spontaneously tend to drain upon saturation of soil with water. Once the drainage is complete, which in a free-draining soil may take 2–3 days, the soil attains what is called a field capacity. At field capacity, the soil holds a maximum amount of water that it can against gravity. The field capacity is generally expressed in g per 100 g oven-dry soil. The gravitational water that is lost in drainage remains unavailable to plants.

Water movement takes place along the water-potential gradient from soil to plants and then transpirational loss from the leaves of the plants. But how is this gradient established? It is the transpirational loss of water that establishes the water-potential gradient—a condition for water movement from soil to roots, to shoots, and to leaves. The water potential in the roots of a plant system is lower than that of the soil. This allows water movement from the soil to the plant roots. Water potential in the xylem of the shoot of the plants is further lowered than that in roots; thus, water moves from roots to shoots and subsequently to the atmosphere via leaves. In this way, water becomes available to the whole plant.

WATER PLANET A CLIMATE-SMART PLANET

The planet's climate change is hidden in what is the largest component of it. That is, water, the most of the planet. The planet's climate solutions are hidden in the largest component of the water. That is, oceans, the most of the water on the water planet. Covering about 71% of the earth's surface, the oceans are home to a vast majority of living things of the planet. The waters outside the oceans and seas—lakes, rivers, rivulets, wetlands, etc.—are all vital sources of climate benevolence on the Living Planet.

Waters of the earth are not just a resource to reckon with. They are a phenomenon in themselves. The hydrological cycle itself serves as a dominant factor of ecological integrity of the planet. The role of water in maintaining the ecological integrity also involves its role in holding the planet's temperatures in an appropriate range. Water is integral to life as well as a medium of life (cells, organisms, ecosystems, the entire biosphere) that resists changes in temperatures. Previously in this chapter, we have discussed that water in organisms' cells acts as a buffer, not allowing temperature changes beyond a range that is attributable to its high specific heat, or heat capacity. Water holds temperatures in a range soothing for the metabolic processes to keep going. In the same way, water in the ecosystems and in the entire biosphere acts as a buffer to resist temperature changes beyond a limit.

The temperature-buffering role of water is on account of its very specific properties vital to life already discussed in this chapter. In our contemporary times, the processes of climate change, via global warming, are gradually strengthening, breaking the barrier of normal ranges of temperature fluctuations. The heat budget of the biosphere is being influenced to the extent that the water capacity to resist temperature change is being gradually overreached.

Imagine what would have been the fate of the Living Planet had water not had the trait of holding great amounts of heat without much change in its temperatures. Water being a dominant regulator of the biosphere's thermodynamics, in fact, has a potential to solve climate crises and transform the planet into a climate-smart planet. As the anthropogenic factors are increasingly coming to the fore as the dominant, rather exclusive, source of climate change, the solutions of the climate crises are also hidden in alternative human management of water resources. Depollution of rivers, ponds, lakes, seas, and oceans will enhance their capacities to regulate increasing temperatures. Availability of water to ecosystems and fields and to all the places where plants naturally grow or can grow with our efforts must be ensured. Availability of water would enhance the photosynthetic efficiency of the earth's ecosystems. As photosynthesis alone can regulate a climate pattern via carbon sequestration, the building up of soil-carbon stocks and regulating the carbon cycle to a great extent, increasing the planet's photosynthetic efficiencies, must be the primacy in our programs, policies, and global agendas. Ecological regeneration of degraded forests, conservation of natural forests, and afforestation would be potent to enhance photosynthetic abilities of the earth up to a natural potential. Anthropogenic actions leading to the amelioration of photosynthesis trigger conservation processes, like water, soil, and biodiversity conservation, which lead to the fortification of ecological processes, including a life-soothing climate. The hope of a climate-smart planet, in fact, lies in the waters of the water planet.

SUMMARY

Most of the earth is water and so is life. Water is both an ingredient in anabolism as well as a product of catabolism. Water is vital in maintaining the thermal budget of the biosphere. The polarity of water makes it an excellent solvent. Distribution of water on Earth is unequivocally uneven in terms of space, time, and type. Availability, acquisition, and conservation of water are important considerations, especially in the context of the organisms in desert, arid, and semiarid environments. How the organisms maintain a water balance is a subject of their water relations. Where there is water, there is life. Water embraces all the properties that are conducive to life.

Water moves down concentration gradients. Whether the organism loses or gains water would depend on the water-concentration gradients between organisms and their environment. The most commonly used measure of water in the air is relative humidity, that is, the water content of air relative to its content at saturation. Hot air has more potential to hold water vapor than cold air. The pressure the water exerts on air saturated with water is the saturation water-vapor pressure. At sea level, the atmospheric pressure averages approximately 760 mm Hg. The international convention to represent water-vapor pressure is pascal (Pa). There is difference between the amount of water the air can potentially hold and the amount of water at a particular point of time. This difference is known as VPD. Water flow from an organism to the atmosphere in terrestrial environments takes place in accordance with the VPD of the air surrounding the organism. The concentration of water is a decisive factor determining the direction of water flow in aquatic environments. Depending on the amount of water per unit volume (or concentration of the water), the aquatic organisms are divided in three categories: (1) Isosmotic (concentration of water in the body fluids and external environment is alike; there is no movement of water between organisms and their environment); (2) Hypoosmotic (concentration of water in body fluids is higher than in the external environment;

organisms tend to lose water to the external environment); and (3) Hyperosmotic (concentration of water in body fluids is lower than the external medium; organisms tend to be water flooded due to movement of external water toward their body fluids). When moving along its concentration gradient, water produces an osmotic pressure, also measured in pascals (Pa).

Water potential, measured in (Pa) and denoted by the symbol ψ (psi), is the capacity of water to do work. Flowing water has its kinetics to do work. The water potential of pure water is set at 0, the highest value of water potential in nature. The water potential of a solution is expressed as: $\Psi = \Psi_{solutes}$. The water potential of the fluids within plant cells is approximately: $\Psi_{plant} = \psi_{solutes} + \psi_{matric}$. Water potential is affected (reduced) by solutes, matric forces, and negative pressure: $\Psi_{plant} = \psi_{solutes} + \psi_{matric} + \psi_{pressure}$. As water moves from the soil to the roots due to $\Psi_{soil} > \psi_{plant}$, it joins a column of water extending from the roots through the xylem (water-conducting cells) to all parts of the shoot of a plant.

There is a definite relationship between root development and water acquisition by the plants. Soil moisture influences root growth of the plants. Root development is also associated with climate. The climate itself determines the availability (or deficiency or unavailability) of water for the plants. Plants occurring in dry climates tend to develop deeper root systems than those in moist climates.

The external environment of a leaf includes cell walls and intercellular spaces, or leaf apoplasts, which are subject to atmospheric pressure, that is, $\Psi_p = 0$. The solute concentration is quite low, contributing to a lower value of Ψ_s. Matric forces exerted by the cell walls, therefore, determine water potential in the apoplast (Ψ_{apo}). Thus, $\Psi_{apo} = \Psi_m$. Because of the thermodynamic equilibrium within a cell, there is much likelihood that the water potential in the cytoplasm (Ψ_{cyt}) equals to that in a vacuole (Ψ_{vac}), that is, $\Psi_{cell} = \Psi_{cyt} = \Psi_{vac}$.

The quantity of water in a particular soil depends on climate and precipitation in excess of evapotranspiration (i.e., P – E). In areas where P > E, water availability for plants is no problem. The greater the difference between P and E, the greater the availability of soil water for the plants. Vegetation in the P < E-ridden areas is less dense, and plant distribution is sparse. Vast patches of the land in extremely dry areas might be devoid of any vegetation. Water movement takes place along the water-potential gradient from soil to plants, and then transpirational loss from the leaves of the plants takes place.

The planet's climate solutions are hidden in water, the largest component of the planet. Waters of the earth are not just a resource to reckon with; they are a phenomenon in themselves. The hydrological cycle itself serves as a dominant factor of ecological integrity of the planet. Waters of the earth also contribute to maintain the ecological integrity by holding the planet's temperatures in an appropriate range, which is attributable to its very high value of specific heat, or heat capacity. Water in the organisms' cells acts as a buffer by maintaining the temperature in a range soothing for the metabolic processes to keep going. In the same way, water in the ecosystems and in the entire biosphere acts as a buffer to resist temperature fluctuations beyond a limit.

REFERENCES

Khatoon, H. and Singh, V. 2017. Effects of water stress on morphological parameters of finger millet (*Eleusine coracana* L.). *Environ. Ecol.* 35 (2D): 1595–1604.

Kramer, P. J. and Currier, H. B. 1950. Water relations of plant cells and tissues. *Annu. Rev. Plant Physiol.* 1: 265–284.

Molles, M. C. 2005. *Ecology: Concepts and Applications*, 3rd edn. Boston, MA: McGraw Hill, pp. 127–128.

Schenk, H. J. and Jackson, R. B. 2002. The global geobiography of roots. *Ecol. Monogr.* 72: 311–328.

Webster, R. and Beckett, P. H. T. 1972. Matric suctions to which soils of South Central England drain. *J. Agric. Sci.* 78: 379–387.

WEBSITES

http://employees.csbsju.edu/SSAUPE/biol327/Lab/water/water-hofler.htm

https://www.reuters.com/investigates/special-report/ocean-shock-warming/

5 Temperature Relations

Why are *Homo sapiens* so concerned with temperature? For us, and all other species, the impact of extreme temperatures can range from discomfort, at a minimum, to extinction.

Manuel C. Molles Jr.

MICROCLIMATIC VARIATIONS AND TEMPERATURES

Plants occupy a microspace in the biosphere: roots up to a few meters into the earth's crust, shoots and canopies up to a few meters into the earth's atmosphere. This very thin "green layer" of the biosphere, however, is critical for the whole life within the biosphere and for the biosphere itself. The space the plants occupy is very reactive due to intensive interactions between the physical environment of the lithosphere, hydrosphere, and atmosphere and living organisms (plants, animals, and microorganisms). The relationships between the physical environment and living organisms are complex and reciprocal, both influencing each other. Because the extent of the physical environment is much more than that of the "green layer," the former determines the latter and the latter influences the former to a certain extent. In other words, the green layer is a "product" of the physical environment, and the physical environment itself is influenced by the green layer.

Among the physical factors determining the status (structure and functioning) of the green layer is a non-terrestrial one that is the most critical one: the light. The light that is incident upon a surface is either direct sunlight or diffused light. The diffused light is the light scattered by particles and droplets in the atmosphere. Because the light is also absorbed by the atmosphere, there is more intense radiation at high altitudes with a relatively greater amount of short-wavelength (high-energy) radiation. It is the light that, upon striking an object, generates heat that is measured in temperature. Temperature, thus, is an outcome of light and varies according to the nature of the object (surface) it strikes. The most conspicuous nature of the surface/object determining its temperature is related to the altitude where it lies or is placed at. The nature or properties of the surfaces in accordance with their smoothness, ruggedness, vegetation, color, etc. also determine the extent of temperature, which is quite different from the gross environment that surrounds them. In other words, a diversity of microenvironments is created due to varying properties of the surfaces and altitudes of the surfaces.

The microenvironments built up through interactions between biotic and abiotic factors operating on a surface/area/site make a climate somewhat different from the general climate. We call this climate a microclimate. The diversity of microclimates within a macroclimate is one of the unique features of our environment. A study of the microclimates as characteristics of the microenvironments is essential to understand the specific responses of various plants growing and performing in these environments.

ATMOSPHERIC TRENDS VIS-À-VIS ALTITUDES

Altitude, being the dominant factor influencing temperature and pressure, determines the characteristics of the atmosphere. Altitude, thus, also becomes the basis of the habitat characteristics and eventually of magnitude and characteristics of life. The atmosphere prevails up to an altitude of 500 km. Not the entire range of the atmosphere accommodates and supports life. Since temperature and pressure are the major determinants of the existence (and non-existence), life remains confined to the lower atmosphere. The heat engine, along with the ocean, distributes heat throughout the globe. This phenomenon, along with several other terrestrial and cosmic factors, drives the climate system of the planet. Climate (the long-term weather pattern) and weather (the short-term climate pattern) are the two aspects arising out of the heat distribution pattern on planet Earth. The atmospheric variables significantly influence weather conditions as well as a climate system in a region. Therefore, all regions and all areas of a region on the planet have diverse weather and climate patterns.

Solar radiation, humidity, and winds are the major variables influencing atmospheric thermodynamics. Variations in the atmospheric pressure determine the characteristics of other atmospheric variables.

From the sun, the star in the Milky Way nearest to Earth, energy travels through space in the form of electromagnetic radiation with a wide range of wavelengths: short to long wavelengths in a continuum called an electromagnetic spectrum (Figure 5.1). The amount of energy in radiation depends on the wavelength of the radiation. The shorter the wavelength the more the amount of energy in the radiation, and vice versa. Hot bodies emit radiation of shorter wavelengths and cool bodies of longer wavelength. The surface temperature of the sun is 5480°C and emits radiation with wavelengths in the range of 0.2 to 3.0 μm. As much as about 95% of the solar energy consists of ultraviolet (UV), visible, and infrared radiation. The rest comprises gamma rays, X-rays, microwaves, and radio waves. Wavelength and frequencies of various types of radiation in the electromagnetic spectrum are shown in Table 5.1.

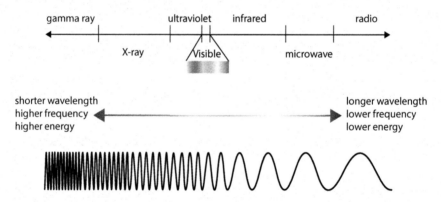

FIGURE 5.1 Electromagnetic spectrum. (Courtesy of NASA's Imagine the Universe, cited in: https://imagine.gsfc.nasa.gov/science/toolbox/emspectrum1.html.)

TABLE 5.1
Wavelength and Frequency of Various Categories of Radiation in the Electromagnetic Spectrum

Category	Range of Wavelength (nm)	Range of Frequencies (Hz)
Gamma rays	<1	>3 × 10^{17}
X-rays	1–10	3 × 10^{16}–3 × 10^{17}
Ultraviolet light	10–400	7.5 × 10^{14}–3 × 10^{16}
Visible light	400–700	4.3 × 10^{14}–7.5 × 10^{14}
Infrared	700–10^{5}	3 × 10^{12}–4.3 × 10^{14}
Microwave	10^{5}–10^{8}	3 × 10^{9}–3 × 10^{12}
Radio wave	>10^{8}	<3 × 10^{9}

The gamma rays constitute extremely powerful radiation of short wavelength; these can penetrate most of the objects and are life-annihilating in their nature. X-rays are almost equally powerful. These two shortwave radiations are absorbed by the gas molecules in the periphery of the earth's atmosphere and, thus, do not directly harm life in the troposphere. The shortwave UV radiation is absorbed mostly in the ozone (O_3) layer of the stratosphere. However, a lesser proportion of this radiation reaches the earth's surface and directly affects life in the troposphere.

Visible radiation (400–700 nm), with much of its proportion entering the earth's atmosphere, reaches the earth's surface. This is the radiation that drives photosynthesis and, hence, serves as an energy source for most of the life on Earth and an exclusive energy source for all life dependent on photosynthesis. This is also the radiation in which we all are able to see everything in this universe. Since the life energy comes to us via photosynthesizers, it is this energy by which we see, hear, feel, taste, smell, and drive all our metabolism and engage in all life activities. The visible radiation on the spectrum is composed of seven colors (VIBGYOR), with violet color on the short wavelength end and the red on the long wavelength end of the color spectrum. The longwave infrared radiation also reaches the earth's surface. It, however, does not drive photosynthesis but is useful for all life in the biosphere. Microwaves and radio waves, which constitute relatively larger wavelengths compared to all other categories of radiation in the spectrum, reach the earth surface without any obstruction (Figure 5.2).

The temperature of the earth's surface and the lower atmosphere is relatively higher than would be expected of a planet like that of ours placed at such a distance from the sun. The main reason of the same is the greenhouse effect exerted by the so-called greenhouse gases (GHGs), which is one of the major features of our biosphere. Carbon dioxide (CO_2) is the major player in the phenomenon of the greenhouse effect. However, nitrous oxide (N_2O), water vapor (H_2O), methane (CH_4), and chlorofluorocarbons (CFCs) are the other gases playing similar roles as CO_2 in retaining heat in the atmosphere. As a result of atmospheric composition and its impact on the thermodynamics of

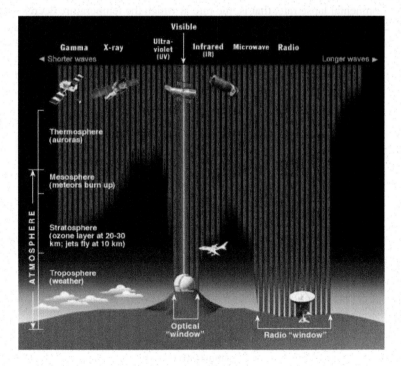

FIGURE 5.2 Absorption pattern of the electromagnetic radiation before reaching the earth's surface. (Courtesy of STScI/JHU/NASA; From https://imagine.gsfc.nasa.gov/science/toolbox/emspectrum1.html.)

the planet, the average surface temperature of Earth is about 15°C. If there were no GHGs in the atmosphere, the planetary surface temperature would be −18°C, that is, 33°C lower than the current surface temperature.

Temperature regimes alter in their behavior in accordance with altitude. On the basis of the altitude–temperature relationship, Earth's atmosphere can be divided into four zones or layers, viz., troposphere, stratosphere, mesosphere, and thermosphere. Temperature behavior in different atmospheric layers determines the temperature of the earth's surface. Sequential changes in these atmospheric layers drive weather and climate of the earth. Patterns of temperature regimes according to altitude determine and substantially influence the structure, composition and functioning of Earth's ecosystems.

TROPOSPHERE

The lowermost atmospheric zone encompasses a temperature regime that decreases with an increase in temperature. It is about 12 km above Earth's surface and extends up to about 1 km (ranging from few meters to 2 km depending on a landform) below Earth's surface where an intensive land–atmosphere interaction takes place. This layer is called as planetary boundary layer. The uppermost layer of the troposphere is known as the tropopause where a decrease in temperature ceases to occur (inversion layer) and makes a boundary between the troposphere and stratosphere. Most of the atmosphere gases are concentrated in this zone. Most of the atmosphere's mass (about 75%) and 99% of the total mass of H_2O and aerosols are confined to the lowest atmospheric zone.

The gaseous composition (N_2 78.08%, O_2 29.95%, Ar 0.93%, CO_2 0.04%, H_2O variable, and other gases in traces) of the troposphere is pretty uniform except for H_2O. The troposphere is the most dramatic zone from the viewpoint of temperature-led physical, chemical, and biological reactions between the atmosphere, hydrosphere, and lithosphere. Water cycle, weather cycle, biological cycles, and most of the nutrient cycles on the planet are accommodated in the troposphere.

Temperature in the troposphere zone decreases with an increase in altitude as does the saturation H_2O. The environmental lapse rate, that is, the rate at which the temperature decreases (calculated as—dT/dz, where T is temperature in Kelvin and z the height in meters) varies according to altitude. It decreases to the extent of 6.5 Kelvin between 0 to 11,000 m. There is no decrease between 11,000 and 20,000 m. From 20,000 to 47,000 m, an increase in temperature is recorded, and then there is no decrease and increase between 47,000 and 51,000 m, and then there is again a temperature decrease between 51,000 and 85,000 m (Figure 5.3).

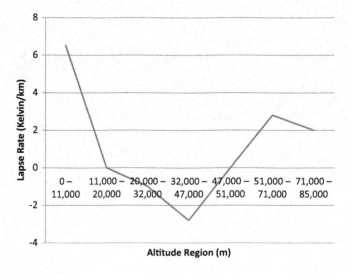

FIGURE 5.3   Environmental lapse rates.

The weight of the air column in the troposphere exerts a pressure on the earth's surface. The pressure at sea level is 1013.2 mb, on average, which decreases logarithmically, roughly 50% for every 5 km of altitude, in accordance with log pressure = 0.06 × altitude in km. As the air warms up due to surface temperature, it expands and becomes less dense. The lighter air rises, and the comparatively cooler and more dense upper air replaces the lower layer; as a result, this process generates convection currents. The lower warmer air, naturally, generates low-pressure areas, and the upper cool air generates high-pressure areas. The air, as a physical rule, blows from a high-pressure area to a low-pressure area. The warmer air holds more H_2O than the cold air. Therefore, when rising up in the atmosphere, the warmer air carries with it certain amounts of H_2O. The amount of H_2O in the atmosphere becomes a measure of humidity. When the air is cool enough to condense the H_2O in the air at the critical temperature, it leads to the formation of clouds, which eventually will precipitate as rain or snow. The clouds play an important role in reflecting back solar radiation and, thus, in controlling the thermal budget of the biosphere to some extent.

The solar energy that reaches the earth's surface accomplishes various dimensions supporting and enhancing life. These include photosynthesis (conversion of CO_2 and H_2O into organic matter and living energy); water evaporation (hydrological cycle); and driving winds, currents, and waves (distribution and regulation of atmospheric heat); and absorption by GHGs (heat retention for maintenance of temperature range conducive to the biosphere). The amount of CO_2 in the atmosphere is critical for the amount of heat energy to be retained. The larger the atmospheric CO_2 amount, the greater the amount of the heat retained. H_2O, N_2O, CH_4, and artificially synthesized CFCs, all behaving as GHGs like CO_2, also significantly contribute to retain the outgoing shortwave radiation. The outgoing shortwave radiation the CO_2 and other GHGs retain phenomenally influences the climate of the planet. The greater the amount of GHGs, the warmer the climate.

STRATOSPHERE

The stratosphere extends from about 12 to 48 km above the earth's surface. The most striking characteristic of this atmospheric zone is the occurrence of the O_3 layer. The O_3 serves to absorb the incoming shortwave UV radiation in the top two-thirds (24–48 km) of the stratospheric zone. The energy of the incoming UV radiation breaks down oxygen molecules (O_2) in the stratosphere into free oxygen atoms (O) which, in turn, unite with other O_2 molecules to produce O_3 ($O_2 + O = O_3$). The O_3 molecules are very effective at absorbing UV rays, and the stratospheric O_3 layer absorbs as much as 97%–99% energy of the incoming UV radiation from about 200–315 nm wavelength.

In the process of UV filtration, breakdown and regeneration of O_3 are in balance. However, due to overwhelming anthropogenic interference, O_3 depletion processes are exacerbating and this, in recent years, has become a worrisome issue. N_2O emitted from the earth's surface and artificially synthesized CFCs deplete the ozone layer, and continuous O_3 depletion would be extremely detrimental to life on Earth.

Warming up of the stratosphere occurs due to absorption of UV radiation. Temperature in this atmospheric zone, therefore, is higher than that in the troposphere and increases with increase in altitude. Air density also goes on decreasing as the temperature increases along the altitude within this zone. The warmer stratosphere acts as a lid on the denser atmosphere, trapping the cooler weather system in the troposphere. The stratosphere zone is, therefore, more stable than the troposphere. Gases and aerosols that make entry into stratosphere due to volcanic eruptions, nuclear explosions, industrial processes (e.g., CFCs), and jet-plane exhausts are trapped there for years. The presence of these pollutants in the stratosphere affects the thermal regulation of the planet. Jet-airplane trails, the contrails, are often visible in the relatively still air in the tropopause— between the troposphere and stratosphere.

Shivaji et al. (2009) found bacterial life in the stratosphere. They isolated three novel bacterial species, namely, *Janibacter hoylei* sp. nov., *Bacillus isronensis* sp. nov., and *Bacillus aryabhattai* sp. nov. at altitudes between 27 and 41 km using cryotubes to collect air samples. Some birds have also been encountered flying above the troposphere. These observations reveal that the stratosphere is a part of the biosphere.

MESOSPHERE

Above the troposphere and the stratosphere lies the mesosphere in the altitude range between 50 and 85 km above the earth's surface. The air in this atmospheric layer is much thinner than in the stratosphere and the troposphere. Much less-dense air and small amounts of O_3 present in this layer are the reason why there is not considerable warming of this atmospheric layer through the incoming radiation. Much radiative cooling takes place in this atmospheric layer. CO_2 molecules absorb heat energy when they bounce off of other molecules. A proportion of the energy CO_2 absorbs in this layer is released as photons in a process called radiative emission. The photons carrying heat travel upward into thermosphere, thus preventing the mesosphere from getting warm and cooling with the increase in altitude until the temperature of the layer in the uppermost region reaches about –100°C. A continuous decrease in temperature along increasing altitude characterizes the extent of the mesosphere between the stratosphere and thermosphere.

The mesosphere is the coldest of all the atmospheric layers. In fact, it is colder than even the polar region of Earth, and even that of the surface of the planet Mars. Before merging into the thermosphere, the second outermost atmospheric layer, the mesosphere carves out its outer layer between 80 and 90 km, known as the mesopause.

THERMOSPHERE

The layer of the earth's atmosphere lying between the mesosphere and the exosphere is known as the thermosphere (Greek: *Thermos* means heat). The thermosphere extends at an altitudinal range from 85 km to between 500 and 1000 km from the earth's surface. It means that the thermosphere is thicker than the lower atmospheric layers. The temperature soars rapidly in the lower thermosphere (between 200 and 300 km) along the altitude, and then it levels off in the upper portion of the atmospheric layer. In the upper portion of this layer, the temperature ranges from about 500°C to about 2000°C. Solar activity phenomenally influences the temperature of the thermosphere. When solar activity is very intense, the upper portion of the thermosphere, the thermopause, expands or "puffs up."

The heat of the thermosphere is actually the heat that won't warm you if you hang out in this part of the atmosphere. It is because there are not enough molecules in this atmosphere that could transfer heat to your body.

Intense and high-energy X-rays and UV radiation, much of which are absorbed in the thermosphere, cause photoionization of the air molecules forming what is also called ionosphere. In the upper thermosphere, atomic nitrogen (N), atomic oxygen (O), and helium (He) are the main air components. The electrically charged particles in this layer enable radio waves to be refracted and received even beyond horizons.

Although the thermosphere is part of the earth's atmosphere, because of very low air density, it is sometimes regarded as part of space. Indeed, space is thought to begin from the altitude of 100 km, a few km from above the mesopause. Both the International Space Station and the Space Shuttle orbit Earth in the upper part of the thermosphere.

The Southern and Northern Lights, popularly known as "Aurora," primarily occur in the thermosphere. Electrons, protons, and other ions—the charged particles present in space—collide with atoms and molecules in the thermosphere at high latitudes, exciting them into higher states of energy. This colorful and enchanting aurol displays occurs due to the shedding of excess energy by those atoms and molecules.

Exosphere

The fifth and the outermost thin atmospheric layer of the earth just above the thermosphere is what is called the exosphere. The exosphere extends up to about 10,000 km from the earth's surface (little less than the width of Earth) and goes on thinning with increasing altitude until it gradually merges with planetary space. The lower boundary where the exosphere meets its next layer toward the earth is called the thermopause. The thermopause tends to increase in size with the increase in the intensity of solar activity. Swelling of the thermosphere, therefore, leads to an increase in the extent of the exosphere. The outer boundary, or edge, of the exosphere is difficult to define. Sometimes it is "demarcated" with the presence of hydrogen as a glow of UV rays, a region called the geocorona. And yet another reason also carries weight: where hydrogen atoms surpass Earth's gravitational force due to the pressure from solar radiation. The distance of the exosphere's edge is quite far away from the earth's surface: 190,000 km. In its extent, the exosphere's top boundary goes up to about halfway to the Moon.

The composition of the exosphere is simple. The atoms in this layer are so far away from each other that a collision among them is a very rare phenomenon. Hydrogen gas is present throughout in this layer. Some amount of He, CO_2, and nascent O also occur but toward the base of the exosphere. The temperature of the exosphere is also difficult to measure because of its thinness. It is quite cool. Gas molecules and atoms in this outermost layer move along "ballistic trajectories." Some faster-moving particles do not return to Earth along its gravitational pull. This is how a small portion of the earth's atmosphere "leaks" into space each year.

The planet Mercury, the Galilean moons of Jupiter, and the Moon, unlike Earth, have an exosphere without a dense atmosphere below it. This is called surface-boundary exosphere. Molecules in the surface-boundary exosphere are ejected on elliptical trajectories before their collision with the surface. Smaller heavenly bodies, including asteroids, do not have a surface-boundary exosphere, so the molecules emitted from the surface of such bodies fly off into space.

PLANT–TEMPERATURE RELATIONS

Plants are not akin to homeothermic animals. They cannot maintain their physiological processes at a constant optimum temperature. Changes in environmental temperatures, which do occur invariably, affect the physiology and growth of plants. However, there is no linear relationship that we can establish between plant processes and environmental temperatures. It is because there is enormous variation in air and soil temperatures. Regular seasonal and diurnal temperature variations, irregular and short-term cloudiness and winds, leaf dimensions (size and shape), leaf position in canopy ("sun" or "shade" leaves), plants' heights above and depths below the soil surface (shoot-root lengths) etc. are the dominant terrestrial factors the temperature of the plants depends upon.

Plant physiology is influenced by unique patterns of temperature fluctuations determined by the previously mentioned factors. The plant-growth rate transforming into plant distributions and ecosystem structure and composition on a long-term basis is influenced by a mix of thermal parameters, such as minimum, maximum, and mean temperatures, and the amount of the accumulated temperature (e.g., degrees hours and/or degrees days) above a threshold during a season, or a year, or a decade.

Apart from the previously mentioned temperature relationships of the plants, their different development stages and different physiological processes may have different temperature optima. In most of the geophytes, for example, temperature optima for flower-meristem induction and early stages of floral organogenesis range from 9°C to 25°C followed by considerably lower temperatures for several weeks that vary between 4°C and 9°C in the autumn season, which is a necessity for stem elongation and anthesis. In order to complete their annual cycle, thus, most of the geophytes require a "warm-cold-warm" sequence of temperatures (Khodorova and Boitel-Conti 2013).

Temperature has been shown to be the most important variable affecting seed germination (Milbau et al. 2009). Seed germination is influenced by temperature on the basis of three prevailing conditions: (1) moisture content, (2) enzyme activity, and (3) hormone production. For their germination, seeds need to imbibe water, which depends on the moisture content of the soil. A substantial increase in temperature increases the evaporation rate decreasing the soil-moisture content, thus affecting the increasing water requirement for the seeds to germinate. Since enzymes responsible for germination and hormones responsible for seedling growth function only in aquatic medium, elevated temperatures will substantially reduce seed germination.

Two plant hormones, viz., abscisic acid and gibberellins, regulate seed germination. While abscisic acid promotes dormancy and inhibits germination, gibberellins advance the process. Environmental cues, primarily the temperature, upregulate the genes that control gibberellins. As a result, dormancy is released, embryo elongates, radical protrudes, and germination occurs (Finch-Savage and Leubner-Metzger 2006). A considerable increase in temperature causes the enzymes to become inactive. Because temperature rise is inevitable in the event of gradual increase in atmospheric CO_2 concentration, the drastic effect on seed germination, especially owing to temperature dependency of enzymes and hormones, would be inevitable.

The characteristic response of plant growth to temperature is owing to the fact that a rise in temperature affects biochemical processes in two ways that are mutually antagonistic:

1. With an increase in plant-cell temperature, the velocity of the movement in all three dimensions of the reacting molecules, viz., rotational, vibrational, and translational, increases, resulting in increased frequency of intermolecular collisions and more rapid reaction rates, and this influence is common to most of the chemical reactions occurring in the plants.
2. With the rise in temperature, enhanced molecular agitation tends to damage the three-dimensional shapes of the enzyme proteins, resulting in reduced enzyme activity and reaction rates.

GROUND COLOR AND TEMPERATURES

Where vegetation does not usually cover the ground, the color of the ground matters in influencing the prevailing temperatures. Arid and semiarid landscapes of the world bearing bare grounds reflect a variety of colors. Many deserts of the world are named after the colors of their grounds: for instance, Karakum (Turkish: black sand) and Kyzylkum (Turkish: Red Sand) in Central Asia, White Sand Desert in the Tularosa Basin of New Mexico, and the likes. The beaches of the world also offer bare grounds with a wide variety of colors. For example, beach areas in New Zealand range from black to white in color. The diversity of colors the beaches offer provides a diversity of microclimates for the organisms to prevail.

White grounds of the landscapes reflect more light, causing the landscapes to warm up to certain degrees. However, the darker grounds reflect comparatively less light and absorb comparatively more light with comparatively greater rise in temperatures. Thus, if the two landscapes with varying colors are exposed to the same microclimate, the rate at which they warm up will be different: light-colored ground will heat up less compared to dark-colored ground. The white- and black-colored grounds would depict the sharpest contrast as the white reflects all wavelengths of visible light, while the black ground absorbs all wavelengths of visible light.

VEGETATION AND TEMPERATURES

Vegetation has a phenomenal impact on the temperatures. Trees intercept the light reaching the ground and thus lower the temperatures. Casting its shadow, a tree creates a microclimate characterized by somewhat lower temperatures than that of the surrounding environment. If there are shrubs in the lower story (below the tree), a decrease in the temperature would be more. Temperature would further decline if there are grasses and herbaceous plants below the trees and the shrubs. If there is lot of plant litter along with trees, shrubs, and herbaceous plants, a decrease in temperatures will still be more. In this way, a multi-story forest with close canopy, floor vegetation, and litter creates its own microclimates significantly varying from the general climate. Canopy cover of the forests is the major regulator of the surrounding temperatures. The denser the canopy cover, the lower the temperatures.

The microclimate created by vegetation creates an appropriate habitat for a number and varieties of living organisms—animals and microorganisms. By changing the macroclimate of a region, the vegetation also helps build up a moisture regime and ameliorate the quality of the soils. There is always more life amidst the environment of the vegetation, in essence.

SLOPE ASPECTS AND TEMPERATURES

The plain landscapes exhibit a more uniform climate than the undulated ones. Slope aspects in the hills, mountains, and valley areas cast their shadows on the land and, thus, influence the temperatures of the area. The shadows the mountains and hill slopes cast are phenomenal in creating a microclimate different from the macroclimate. The longer the period an area of the land is exposed to shadows cast by mountains or hill slopes the greater the decrease in temperature. In the Northern Hemisphere of Earth, the shadowed areas are on north-oriented areas or the northern aspect, while in the Southern Hemisphere the shadow is cast over the southern aspect. It is because in the Northern Hemisphere the northern aspect faces away from the equator while in the Southern Hemisphere the southern aspect faces away from the equator.

In the Himalayan mountains, for example, the north-oriented slope remains exposed to solar radiation for a lesser period, or shadowed for a longer period, than the south-oriented slope. The northern slope, thus, is cooler than the southern slope. The north-west slope receives minimum solar radiation and forms the coolest places. The south-east slope, on the other hand, remains exposed to sunlight for longest period and, therefore, forms the warmest places. Further, the north-oriented slopes contain more moisture while the south-oriented are the drier ones.

Impact of the slope aspects on prevailing temperatures in the mountains and hills, in fact, is phenomenal in creating conspicuous microclimates. Data from sand dunes in the Negev Desert documented by Molles (2005) revealed a temperature difference from 7.8°C to 9.5°C at midday in the winter and from 1.8°C to 2.5°C at midday in the summer.

It is due to a more appropriate climate that the north-oriented slopes support dense vegetation and have thick soil cover. The southern slopes, on the other hand, have shallow soil cover and do not support an appreciable variety of plants to thrive. In the Himalayan region, the normal landscape depicts dense oak forests on northern slopes and pine monocultures or bare patches on southern aspects. The slope-orientation coupled with altitudes in the mountains also offers a variety of distinct ecological niches that farmers use for producing different crops and derive an advantage of specific products and functions emanating from these specific ecological niches.

AQUATIC ENVIRONMENTS' TEMPERATURES

Fluctuations in air temperature are very high and frequent. Water environments, however, are quite stable to temperature fluctuations. The more the fluctuations in temperatures in an environment the higher the degree of turbulence, and vice a versa. Thus, air is more turbulent in response to temperature fluctuations than water. The higher degree of thermal stability of water is partly owing to its very high specific heat—the capacity of absorbing heat without undergoing a rise in temperature. The capacity of water to absorb heat without registering a rise in temperature is 3000 times greater than that of the equal volume of air. It takes 1 cal of energy for 1 cm^3 of water to heat it to increase its temperature by 1°C. An equal volume of air for raising an equal temperature requires only 0.0003 cal.

The second reason for the thermal stability of water is its capacity for absorbing large amounts of heat as it evaporates—that is, the latent heat of vaporization. It takes as many as 580 cal at 35°C for 1 g of water from its surroundings to get converted into a gaseous state. At 22°C it requires 584 cal for the same purpose. In other words, we can infer that it takes little water to impact the cooling effect to air or living matter.

The third reason for the extraordinary thermal stability of water is owing to its latent heat of fusion, that is, it yields heat energy upon changing from its liquid state to the solid state. One gram of water, in the process of its freezing, gives off 80 cal of heat to its surrounding environment. As the water leaves its liquid state to be incorporated into the crystalline latticework of ice, the energy of the motion required by molecules exits into its surrounding environment. In other words, in a water body 1 g of water gives off sufficient energy to heat 80 g of water by 1°C, thus retarding further cooling.

Extreme thermal stability is demonstrated by a large water body, such as an open ocean, which records only a temperature fluctuation of about 1°C. Even the large streams record a very small range of fluctuations compared to their surrounding air and terrestrial environment.

In addition to the physical properties of water, viz., high values of specific heat, latent heat of evaporation, and latent heat of fusion, riparian vegetation (the vegetation that grows along rivers and streams) also contributes to influence the temperature of aquatic environments the same way it does in case of terrestrial environments as discussed elsewhere. The riparian vegetation casts its shadows on the streams and insulates the surrounding environment, thus further reducing temperature fluctuations in the lotic ecosystems.

PHOTOSYNTHESIS IN EXTREME TEMPERATURES

Photosynthesis is the most fundamental ability of green plants, and this becomes the basis of survival, sustenance, and evolution of most of the heterotrophic life on Earth. At what efficiency or rate the green plants photosynthesize, however, depends on a number of factors, temperature being one of them. Extreme temperatures generally reduce the rates of photosynthesis. Under ideal conditions set by operating environmental factors (e.g., intensity of solar radiation, moisture content of the environment, and soil health), a plant photosynthesizes at a maximum rate within a narrow range of temperatures. The temperatures at which photosynthesis reaches at its peak would vary according to ecosystem types. In extreme cold, plants attain maximum photosynthetic rates at lower temperatures while in extremely hot climates they would attain the peak of photosynthesis at high temperatures. For example, based on the review of an experiment, Molles (2005) explains that a moss (*Pleurozium schreberi*) from the boreal forest and a shrub (*Atriplex lentiformis*) from a desert ecosystem photosynthesize at maximum rates at 15°C and 44°C, respectively. These temperatures have a huge difference to drive photosynthesis at its maximum in two different ecosystem types. Inherent with the ecosystem type is the genetic makeup of the plants thriving in that environment. At 15°C at which *Pleurozium schreberi* photosynthesizes at its maximum rate, *Atriplex lentiformis* photosynthesizes at only 25% of its maximum photosynthetic rate. This difference in the most fundamental characteristic of the two species occupying different ecosystems is substantial.

Responses of the plants to temperature reflect the short-term physiological adjustments, that is, acclimation. Acclimation, in response to temperature, only involves physiological, not genetic, changes, and it is generally reversible with changes in the environmental conditions. If two clones of the same plant developed through vegetative growth are further grown in two different environments characterized by different temperature ranges, their photosynthetic rates would reach the peak at different temperatures. For example, as Molles (2005) documents, the two genetically similar clones of the desert shrub *Atriplex lentiformis* when grown in cool and hot environments, their photosynthetic rates attained peaks at 32°C and 40°C, respectively. This suggests that acclimation by a plant species may shift its optimal temperature for maximizing its photosynthetic efficiency in consonance with seasons with varying environmental temperatures.

SOLAR ENERGY–TEMPERATURE INTERACTION

In food crop plants in cultivated areas, dry-matter yield, which is the integral of net CO_2 assimilation over the total crop-growing season, is of key importance. The environmental factors that often, rather inevitably, change during a growing season include temperatures of air and soil, which influence crop yields/plant biomass. These variables might be used to derive advantages in increasing the dry-matter yield of the crops.

Soil-surface temperatures at a particular location are strongly influenced by the amount of radiant energy received. The amount of radiation depends on the intensity and duration of solar radiation falling on a location inhabiting the plants. The total amount of radiation received influences the air and the soil temperatures, both together. The earth's surface, however, maintains some amount of residual energy usable in taking time in warming up or cooling down subject to changes in the incoming solar energy. This helps create a lag time at the location between: (1) the highest energy and the highest temperature levels, or (2) the lowest energy and lowest temperature levels. It is because of this lag time that at a particular temperature the radiation energy level is higher in summer than in winter. Again, as the plant biomass accumulation rates or plant growth rates are closely correlated with the solar-energy interception, the higher solar energy level will create a potential for highest biomass or yields when there is an occurrence of the critical leaf-area index (LAI). If a crop grows a maximum leaf area during the availability of the highest energy level, it will derive advantage to fortify biomass accumulation rates and gain higher yields.

TEMPERATURE REGULATION BY PLANTS

Maintaining a balance of heat gain against heat loss is an inherent feature of all the living organisms. Plants, like ectothermal animals, depend on external sources of energy for thermal regulation. Though the ectothermal animals, unlike plants, are capable of maintaining mobility, both solve their energy-balance-related problems in a similar manner. The environments that pose challenges to plant species, that is, relatively harsher environments (tundra and deserts, for example), are more appropriate for the studies relating to thermal regulation by plants.

In order to exchange heat energy with their environments, plants often use their morphological behaviors. One of the well-known examples is the orientation of sunflower toward the sun (Figure 5.4). While in animals many modes of behavior can be observed, morphological behavior in plants are not very conspicuous.

Arctic and Alpine Plants

For an understanding of the heat regulation in plants, we can use the following equation:

$$H_s = H_{cd} \pm H_{cv} \pm H_r$$

where H_s is the total heat stored in the body of the plants; H_{cd} is the heat gained or lost through conduction; H_{cv} is the heat gained or lost by convection; and H_r is the heat gained or lost through electromagnetic radiation.

The natural challenge before the arctic and alpine plants is to avoid excessive cooling and staying warm. For this to happen, they have two options, viz., reduce H_{cv} and/or increase H_r. Most of the plants in extremely low temperatures have evolved both the characteristics and, therefore, have abilities to maintain temperatures higher than that of the air. Nature has especially been benevolent for the arctic and alpine plants by imparting dark pigments that help the plants absorb more solar energy, thus enhancing radiant heat, H_r. Some plants in these extremely cold ecosystems, for example, *Dryas integrifolia* (Figure 5.5), orient their leaves and flowers perpendicular to the sun's rays, and thus, increase H_r gain. And yet, some plants in the extremely

FIGURE 5.4 The sunflower is a unique oft-quoted example of orientation of its flower in the direction of the sun, a behavior of the plant to strike a heat balance. (Courtesy of pexels-photo-2454515.jpg.)

FIGURE 5.5 Sun-tracking behavior of the arctic plant *Dryas integrifolia*: The plant orients its leaves and flowers perpendicular to the sun's rays, enhancing radiative heat gain, H_r. (Courtesy of mountain-avens-1147499_1280.jpg.)

cold climate develop "cushions" that touch the ground and help them gain infrared radiation, increasing H_r. Some plants gain heat through conduction (H_{cd}) by coming in contact with a warmer substrate.

The cushion growth in the plants of extremely cold environments also helps reduce heat loss through convection (H_{cv}). It occurs in two ways: (1) growing close to ground, it helps reduce air movement; and (2) reducing the ratio of the surface area to volume helps slow down the movement of air inside the plants. As a result of these processes, the cushion-bearing plants are often warmer than the surrounding air and also than the plants bearing other morphological forms.

TROPICAL ALPINE PLANTS

Tropical alpines are altogether distinctive from temperate alpines and tropical environments. These habitats are characterized by "winter" nights and "summer" days. The drastic fluctuation in daytime and nighttime temperatures is something rare that the tropical alpine plants have to experience and survive in. Natural selection of plants in these habitats includes the giant rosette plants, a very rare example of convergence (Figure 5.6).

The giant rosette form helps protect the plants against day–night temperature fluctuations. Dead leaves are not shed by the rosette plants. Instead, they insulate the stem and prevent heat loss through convection, H_{cv}, a process that prevents the plants from freezing temperatures at nighttime. Most of the species in the tropical-alpine habitats bear leaves covered with 2–3 mm thick pubescence— serving like a plant fur—that creates dead air space above the leaf surface, which helps the plants save convective heat loss, H_{cv}, in the cool alpine environment. One of the interesting characteristics of the rosette plants is the secretion of copious amounts of fluids within the rosette structures and in the hollow inflorescence. The watery fluid, owing to its higher specific heat, retains heat in it during daytime that imparts warmth to the plants (H_s) and helps the plants not to reach freezing temperatures at night.

Some of the species in the tropical-alpine environments bear leaves like parabolic structures, which help the plants improve radiative heating, H_r, especially of apical buds and expanding leaves. Some of the plants in this harsh habitat close apical buds at night, imparting a protective cover against freezing temperatures at night.

FIGURE 5.6   Rosette of a tropical-alpine plant secretes copious amount of fluids within the rosettes, thus increasing the capacity of the plant to store heat. (Courtesy of rojnik-3652637_1280.jpg.)

Desert Plants

Desert plants occur in hot environments. Therefore, they need to avoid overheating caused by high temperatures. Reducing storage heat (H_s) is a challenge the desert plants have to face. To meet this challenge, the plants have the capability of using three options:

1. Decreasing overheating by conduction, H_{cd};
2. Reducing radiative heating, H_r; and
3. Increasing rates of convective cooling, H_{cv}.

Unlike the arctic and alpine plants that bear leaves creeping close to the ground surface, the desert plants bear foliage quite above the earth surface to avoid or minimize heat gain through conduction, H_{cd}. A wide ratio of leaf surface area to volume in many desert plants contributes to enhance cooling through convective heat loss, H_{cv}. Many desert plants have reflective surface that facilitates the reduction of radiative heat gain, H_r.

The visible light constitutes a little less than 50% of the total solar radiation reaching the earth's surface. A thick layer of white hair on the shoot of many desert plants (Figure 5.7) reflects visible light, facilitating the reduction in radiative heating, H_r. For example, highly pubescent summer leaves in *Encelia farinosa*, a desert plant species, can reflect as much as 46% of solar radiation, restricting heat gain by radiation, H_r.

A change in the orientation of leaves and stems also helps some of the desert species to control heat gain by radiation, H_r. For example, some plants can orient their leaves parallel to sunrays by closing them when the sun is intense at midday. This orientation effectively controls H_r values maintained by some plants in the hot environments.

Temperature Regulation by Thermogenic Plants

Almost all the plants on Earth, except those in the family Araceae, belong to the category of poikilothermic ectotherms. The plants of the Araceae family distinctively use their metabolic energy for warming up of their flowers. Most of the members of this family are tropical. However, those

FIGURE 5.7 A thick layer of white hair on the shoot of many desert plants reflects visible light, facilitating a reduction in radiative heating, H_r. (Courtesy of newmexico-1007939_1280.jpg.)

prevailing in temperate environments are capable of protecting their flowers (and inflorescences) from getting frozen and for attracting pollinators, a process accomplished by utilizing metabolic energy in elevating floral temperatures.

Among the thermogenic plants, the eastern skunk cabbage, *Symplocarpus foetidus*, is one of the extensively studied plants. This species inhabits the deciduous forests of the eastern part of North America from February to March when air temperature ranges between −15°C and +15°C. The lower air temperatures, however, are overcome by the inflorescence of the plant by means of generating heat endogenously, via metabolic processes. As a result, the inflorescence of the plant (weighing 2–9 g) maintains a temperature between 15°C and 35°C. This temperature is high enough to melt the surrounding snow and save the plant, especially the reproductive organs, from cold stress. The inflorescence of *Symplocarpus foetidus*, thus, functions like an endothermal organ to hold the plant in comfortable thermal zone, but only for 14 days.

How is *Symplocarpus foetidus* (Figure 5.8) specialized in the generation of heat and maintaining warmth of its inflorescence? The story begins with a large root system the plant possesses. Roots of the plant are quite a rich repository of starch. A substantial fraction of the starch is translocated to the inflorescence. In the inflorescence, heat generation takes place upon carbohydrate metabolism at a faster pace, generating quite large amounts of heat. Heat generation is used in warming up of the inflorescence. This process not only gives a warming effect to the inflorescence but also helps attract pollinators. In addition, flowers of the plant also give off a sweetish scent that also promotes attraction for pollinators to visit the flowers.

In the process of temperature regulation, *Symplocarpus foetidus* has to maintain unusually high rates of respiration, as high as that of a mammal of similar size (Molles 2005). This is one of the unique physiological features associated with this species. Metabolic rates of the plant increase with a decrease in atmospheric temperature; the lower the air temperature the higher the metabolic rates. This physiological characteristic of the species helps the plants adjust temperature vis-à-vis atmospheric temperatures. Despite substantial variations in atmospheric temperature, the inflorescence of *Symplocarpus foetidus* maintains uniform temperature as per its physiological needs.

FIGURE 5.8 *Symplocarpus foetidus*, popularly known as eastern skunk cabbage, is an endothermic species regulating temperature through metabolism. (Courtesy of skunk-cabbage-1368182_1280.jpg.)

ADAPTATION TO HARSH ENVIRONMENTS

Plants have been thriving almost everywhere on Earth: in the deserts that seldom experience rains, in the snows and icy environments where winters are long and temperatures extremely low, on the floor of multi-story forests where sunrays never reach, and many more types of unusual environmental conditions. A variety of plants in a variety of harsh environments can survive and complete their life cycles because they have evolved certain specific characteristics. Mechanisms associated with specific features that help the plants survive in odd environmental conditions are called adaptations.

Adaptations among organisms are the attributes of natural evolution. Adaptations are inherited traits, not a learned behavior. Organisms come into existence, grow, reproduce, and develop natural behavioral traits amidst a specific set of environmental factors. Environmental conditions are unequivocally dynamic, never uniform. Environmental conditions are especially sensitive to temperatures. Temperature, a measure of heat content of physical environment or of organisms, itself is a factor of so many environmental factors.

Plants growing in deserts often have to encounter the harshest environmental conditions. Certain species, however, survive comfortably even in the deserts that experience only 5.0 cm or even less rainfall annually. Barrel cactus, well adapted to the Sonoran desert stretched from Southern California to Western Arizona, is one of the unique examples.

The barrel cactus is globe shaped when young (Figure 5.9). It has small roots measuring approximately 7.5 cm attaching the plant to the dry desert soil. Scant rains do not help the soil absorb enough water. The shallow roots absorb water quickly from the upper layer of the soil just after rain fall. In the event of a prolonged dry spell, fine ends of the shallow roots fall off. This helps the plant lose water from the plant body into soil. The fleshy stem covered with fine thorns also preserves water. Even strong dry winds cannot evaporate water from the surface of the plants.

In extremely cold environments, plants modify their morphological and anatomical structures and alter physiological traits so as to (1) gain maximum possible heat from their surroundings, (2) lose minimum heat into the environment, (3) maximize the photosynthetic efficiency within a limited period of exposure to solar radiation, (4) efficiently utilize mineral resources from their ecosystems, and (5) maintain optimum growth within their habitats. These aspects of the plants vis-à-vis extreme environments have been discussed at length in Chapters 7 and 8.

FIGURE 5.9   The barrel cactus well-adapted to harsh desert environments. (Courtesy of Lynanne Gelinas from FreeImages; From https://www.freeimages.com/photo/barrel-cactus-1383975.)

PHYSIOLOGICAL STRATEGIES FOR COOLING THE EARTH

There could be a 100,000 ways of cooling the earth, for example, by minimizing GHG emissions, increasing our dependence on clean energy, planting trees, minimum or no exploitation of natural resources, changing our lifestyles, and the like. Our purpose in this chapter is not to discuss some aspects of reversing soaring temperatures the planet is in the clutches of. We shall look into what could be the botanical strategies using some principles worth addressing the soaring temperatures.

The first and the foremost cause of the soaring temperatures of the planet is the disturbance in the carbon cycle. Discovered by Joseph Priestley and Antoine Lavoisier and popularized by Humphrey Davy, the carbon cycle is an active and perfect cycle with atmosphere, hydrosphere, pedosphere, geosphere, and biosphere as the components of carbon exchange. In its atmospheric pool, the carbon (in oxidized form, CO_2) exists only in traces (0.041%). Nevertheless, this is vital for the structure of the biosphere (all living organisms and the complex ecological processes) on the Living Planet. In all the components, except the biosphere, the carbon prevails in inorganic form, whereas in the biosphere it prevails in its organic form—a major constituent of biosphere structure.

Carbon is not a problem. It is a solution. It is a solution because the story of life begins with carbon. Carbon fixes into life and life comes into existence. It is a major constituent of all the biomolecules structuring life and functioning within life. Carbon in its "final pool"—the atmosphere—exists in inorganic form, and in life, in its organic form. Dramatic conversion of carbon from its inorganic form to organic form is the "script" of life that the carbon writes. This conversion requires carbon to be bonded with water in the presence of the energy of light. However, the chemical reaction converting inorganic carbon (in its oxidized form) into the organic form occurs within the chloroplast. The chloroplast serves as a "gateway" for the inorganic carbon to transform into organic carbon. The process we are quite familiar with is photosynthesis. Photosynthesis, in essence, is the phenomenon of generating organic life out of the inorganic environment.

The thermodynamics of the organic world is also determined and controlled by living organisms in the biosphere. The living organisms dominant in determining global temperature are the ones that have the largest proportion of solar energy transformed into the "living" (biochemical) energy, a process that occurs due to photosynthesis. In other words, the larger the proportion of the solar radiation transforming into "living" energy through photosynthesis, the lesser the rise in the earth's temperature. The "living" energy from the photosynthesizers flows into all organisms in the biosphere. Thus, the larger the proportion of solar energy falling on the earth flowing through living organisms (the entire biodiversity), the cooler the earth.

Carbon is also present in ecosystems in organic form outside life—in soil, water, dead tissues of plants and animals. The organic carbon outside life comes through life only: through photosynthesis and chemosynthesis. Since plants are at the heart of our discussion, we shall hold only photosynthesis in our focus. A philosophical understanding of photosynthesis helps us understand the very nature of the ongoing climate change driven by CO_2. Let us understand how.

Photosynthesis removes carbon from the atmosphere and fixes it into life. Fixed into life in organic form, the carbon keeps flowing through living vessels (biodiversity of the earth). When flowing through life, the energy of solar radiation does not end up in warming the earth surface. The proportion of energy that flows through life (all living beings) avoids warming up the earth and gets channelized into the climate-build up phenomenon.

The phenomenon of photosynthesis also combats entropy and strikes an ecological balance. This contribution, in addition, involves climate-change mitigation by removing excess of CO_2 from the atmosphere and capturing a greater proportion of solar energy within life. The shade of trees brings the atmospheric temperature down substantially. There can be an 18°C to 24°C decline in the environmental temperature if there is a cover of trees in a town or city area.

A very large proportion of the solar energy reaching the earth's surface is converted into heat, which is trapped by GHGs leading to the phenomenon of global warming and consequent climate change. If the largest possible proportion of the solar radiation the earth receives is transformed into biochemical energy in the plant biomass, excessive heating of the earth's surface will be avoided. In the process of the conversion of solar energy into biochemical energy, the steady removal of CO_2 from the atmosphere also takes place, which means declined retention of heat by the major GHG, the CO_2. This is precisely the phenomenon involved in our strategy of cooling the earth.

Our well-planned efforts focusing on boosting photosynthesis (with subsequent CO_2 sequestration and climate cooling) to its maximum efficiency may be effectively executed by maximizing solar-energy utilization. Some of the strategies based on maximizing the utilization of solar energy, also applicable in agricultural management, are outlined in Table 5.2. This, in turn, will be realized in enhanced carbon sequestration, increased biomass production, and/or increased yields from the cultivated food crops, increased level of organic matter in the soil (amelioration of soil fertility), and a fortified ecosystem in functioning and productivity. All these changes initiated by amelioration in plant physiology will be phenomenal in addressing elevated temperatures and regulate climate patterns.

Management of the plant physiological processes must be coupled with ecological regeneration of existing forest areas, large-scale afforestation, soil and water conservation, biodiversity

TABLE 5.2
Strategies for Maximizing Solar-Energy Utilization

Strategic Step	Operationalizaion	Implications for Maximizing Photosynthetic Rates/Carbon Sequestration
Leaf-Area Duration Integration of leaf-area index (LAI) with leaf-area duration (LAD)	• Choosing dense canopy trees for plantation/allowing to grow in a forest • Multi-story forests allowing maximum proportion of light to be utilized for biomass synthesis • Transforming croplands into agroforestry systems • Evolving cropping systems and patterns involving C_4 plants	• Utilization of light on area of photosynthesizing tissue and time over time • Allowing interception of solar radiation on photosynthesizing parts over longer periods of time
Solar Energy—Temperature Interaction Net CO_2 assimilation over the total growing season; taking advantage of increased seasonal temperature	• Planting crops/trees that reach critical LAI at the time of temperature increase in a season • Selection/breeding of the plants assuming critical LAI in the warmer period of a crop season • Managing diversity of the plants including those bearing dense canopy (photosynthesizing area) during warmer periods	• Deriving advantage from the elevated temperatures out of annual weather cycles • Harvesting more yields per unit area of cultivated lands
Life Period Optimal for Seed/Fruit Production Longer period of maximum net CO_2 assimilation per unit of soil surface; seed production as a function of the vegetation period; the higher the total biomass production per unit area the more the production of seeds/fruits/ valuable vegetative parts	• Cultivation of late-maturing varieties of a crop in temperate zones • Use of varieties with very long vegetative periods • Breeding of the plants taking advantage of longer photo-periods • Manage natural forests involving evergreen trees/plants	• High rates of seed production over relatively short periods of time in temperate zones • Crops (sugarcane, cassava, forages etc.) that produce vegetative products perform well in the tropics • Forests inhabiting evergreen plants have higher rates of carbon sequestration

management, soil-fertility measures, and sustainable agricultural development. Plant physiology can be used as an effective climate-management strategy when it is applied along with ecologically regenerative and conservation-oriented natural resource management.

SUMMARY

The space that plants occupy is very reactive due to intensive interactions between the physical environment of the lithosphere, hydrosphere, and atmosphere and living organisms (plants, animals, and microorganisms). The microenvironments built up through interactions between biotic and abiotic factors operating on a surface/area/site make up a climate somewhat different from the general climate we call a microclimate. The diversity of microclimates within a macroclimate is one of the unique features of the global environment.

Altitude, being the dominant factor influencing temperature and pressure, determines characteristics of the atmosphere. Altitude, thus, also becomes the basis of a habitat's characteristics and eventually of magnitude and specific features of the life. The temperature of the earth's surface and of the lower atmosphere is relatively higher than would be expected of a planet like that of ours placed at such distance from the sun. The main reason of the same is the greenhouse effect exerted by the so-called GHGs, which is one of the major features of our biosphere. Temperature regimes alter in their behavior in accordance with altitude. Temperature behavior in different atmospheric layers determines the temperature of the earth's surface. Sequential changes in these atmospheric layers drive weather and climate.

Changes in environmental temperatures, which do occur invariably, affect physiology and growth of the plants. Regular seasonal and diurnal temperature variations, irregular and short-term cloudiness and winds, leaf dimensions (size and shape), leaf position in the canopy ("sun" or "shade" leaves), plants' heights above and depths below the soil surface (shoot-root lengths) etc. are the dominant terrestrial factors that the temperature of the plants depend upon.

Plant physiology is markedly influenced by unique patterns of temperature fluctuations determined by the previously mentioned factors. Temperature has been shown to be the most important variable affecting seed germination. Plant-growth rates transforming into plant distributions and ecosystem structure and composition on a long-term basis are manifestly influenced by a mix of thermal parameters, such as minimum, maximum, and mean temperatures and the amount of the accumulated temperature (e.g., degrees hours and/or degrees days) above a threshold during a season or a year or a decade.

Atmospheric temperature is invariably affected by the color of the ground (where there is no vegetation), vegetation cover, and slope aspect (in mountain ecosystems). Water environments, however, are quite stable to temperature fluctuations.

Extreme temperatures generally reduce the rates of photosynthesis. The temperatures at which photosynthesis reaches its peak would vary according to ecosystem types. In extreme cold, plants attain maximum photosynthetic rates at lower temperatures, while in extremely hot climates they would attain the peak of photosynthesis at high temperatures. Responses of the plants to temperature reflect the short-term physiological adjustments, that is, acclimation. Acclimation, in response to temperature, only involves physiological, not genetic, changes, and it is generally reversible with changes in the environmental conditions.

The environmental factors that often, rather inevitably, change during a growing season include temperatures of air and soil, which influence crop yields/plant biomass. These variables might be used to derive an advantage in increasing dry-matter yield of the crops.

Soil-surface temperatures at a particular location are strongly influenced by the amount of radiant energy received. The amount of radiation depends on the intensity and duration of solar radiation falling on a location inhabiting the plants. The total amount of radiation received influences the air and the soil temperatures, both together. The earth's surface, however, maintains some amount of residual energy usable in taking time in warming up or cooling down subject to changes in the incoming solar energy.

Plants, like ectothermal animals, depend on external sources of energy for thermal regulation. In order to exchange heat energy with their environments, plants often use their morphological behaviors. For an understanding of the heat regulation in plants, we can use the following equation: $H_s = H_{cd} \pm H_{cv} \pm H_r$ (where H_s is the total heat stored in the body of the plants; H_{cd} is the heat gained or lost through conduction; H_{cv} is the heat gained or lost by convection; and H_r is the heat gained or lost through electromagnetic radiation).

The natural challenge the arctic and alpine plants face is to avoid excessive cooling and staying warm. For this to happen, they have two options, viz., reduce H_{cv} and/or increase H_r. Tropical alpines are altogether distinctive from temperate alpines and tropical environments. These habitats are characterized by "winter" nights and "summer" days. Drastic fluctuations in daytime and nighttime temperatures is something rare that the tropical alpine plants have to experience and survive in. The natural selection of plants in these habitats includes the giant rosette plants, a very rare example of convergence. Some of the species in the tropical-alpine environments bear leaves like parabolic structures that help the plants improve radiative heating, H_r, especially of apical buds and expanding leaves. Desert plants need to avoid overheating caused by high temperatures. Reducing storage heat (H_s) is a challenge the desert plants have to face. To meet this challenge, the plants have the capability of using three options, viz.: (1) decreasing overheating by conduction, H_{cd}; (2) reducing radiative heating, H_r; and (3) increasing rates of convective cooling, H_{cv}.

Almost all the plants on Earth, except those in the family Araceae, belong to the category of poikilothermic ectotherms. The plants of the Araceae family distinctively use their metabolic energy for warming up their flowers. Most of the members of this family are tropical. However, those prevailing in temperate environments are capable of protecting their flowers (and inflorescences) from getting frozen and for attracting pollinators, a process accomplished by utilizing metabolic energy in elevating floral temperatures.

Adaptations are the inherited traits, not a learned behavior, and among organisms they are the attributes of natural evolution. The first and the foremost cause of the soaring temperatures of the planet is the disturbance in carbon cycle. Applications of the plant physiology tools can help increase photosynthetic efficiencies of the existing terrestrial and aquatic ecosystems that, in turn, would lead to an enhanced rate of carbon sequestration. Plant physiology turns carbon from a problem to a solution. Photosynthesis removes carbon from the atmosphere and fixes it into life. Fixed into life in organic form, the carbon keeps flowing through living vessels (biodiversity of the earth). When flowing through life, the energy of solar radiation does not end up in warming up the earth's surface. The proportion of energy that flows through life (all living beings) combats entropy and avoids warming up the earth.

The principles of plant physiology can be used in formulating an effective climate-management strategy for ecologically regenerative and conservation-oriented natural-resource management. Protection and conservation of existing forest areas, large-scale afforestation, soil and water conservation, biodiversity enrichment, soil-fertility measures, sustainable agricultural development, and our eco-centric lifestyles are pivotal in creating ecstatic pathways to a climate-smart planet.

REFERENCES

Finch-Savage, W. E. and Leubner-Metzger, G. 2006. Seed dormancy and the control of germination. *New Phytol.* 171. doi:10.1111/j.1469-8137.2006.01787.x.

Khodorova, N. V. and Boitel-Conti, M. 2013. The role of temperature in the growth and flowering of geophytes. *Plants* 2: 699–711. doi:10.3390/plants2040699.

Milbau, A., Graae, B. J., Shevtsova, A. and Nijs, I. 2009. Effects of warmer climate on seed germination in the subarctic. *Ann. Bot.* 104. doi:10.1093/aob/mcp117.

Molles, M. C. 2005. *Ecology: Concepts and Applications*. Boston, MA: McGraw Hill. 622 p.

Shivaji, S., Chaturvedi, P., Begum, Z., Pindi, P. V., Manorama, R., Padmanaban, D. A., Souche, Y. S. et al. 2009. *Janibacter hoylei* sp. nov., *Bacillus isronensis* sp. nov. and *Bacillus aryabhattai* sp. nov. isolated from cryotubes used for collecting from the upper atmosphere. *Int. J. Syst. Evol. Microbiol.* 59: 2977–2986.

WEBSITES

http://www.alaskawildflowers.us/Kingdom/Plantae/Magnoliophyta/Magnoliopsida/Rosaceae/Dryas_integri-folia/Integrifolia_02.html

https://calscape.org/Encelia-farinosa-(Brittlebush)

https://earthobservatory.nasa.gov/features/LAI/LAI2.php

https://imagine.gsfc.nasa.gov/science/toolbox/emspectrum1.html

https://www.freeimages.com/photo/barrel-cactus-1383975

https://www.google.com/search?q=Rosettes+of+alpine+plants&tbm=isch&source

https://www.needpix.com/photo/1691093/rojnik-plant-rockery-nature-garden

https://www.needpix.com/photo/484627/new-mexico-desert-flowers-southwest-desert-wild-sunflowers-des-ert-brush-juniper-wildflowers-view-natural-beauty

https://www.needpix.com/photo/download/626096/skunk-cabbage-marsh-plant-free-pictures-free-photos-free-images-royalty-free

https://www.needpix.com/search/Dryas%20integrifolia

https://www.wildflower.org/gallery/result.php?id_image=45383

6 Allelochemical Relations

Whenever a plant evolves to create an allelopathic chemical, its neighbors will evolve defenses; it is an arms race.

Kirsten Findlay

COMPETITION AMONG PLANTS

Life competes for life. All plants compete for their better survival, for realizing potential growth, and for their sustainable existence. They compete with their own species, with other plant species, and also with the organisms outside the Kingdom Plantae. Every plant in the neighborhood of a plant is a competitor. There is hardly a case that a plant is not affected in one way or the other by a plant in its proximity. Competitions are not just only due to limitations of environmental resources (space, nutrients, water, etc.) but also owing to plants' exerting negative influences on each other following various interesting mechanisms. Competition among plants, in essence, is ubiquitous in its influence.

The term "competition" may mean something different to different ecologists. The role of competition, however, remains a mystery. Competition in life is not just for "survival of the fittest" as Darwin would have claimed. A natural process, as per the designs of natural evolution, favors communities, not just individuals. Competition among organisms appears to be a kind of "competitive mutualism." The overall purpose of competition, therefore, appears to be life enhancing.

DEFINING COMPETITION

Competition is an interaction between individuals and populations that is negative for both (Weiner 1994). Interaction between two inter- or intraspecies leading to a negative impact on each other might be physical also. Some plants, on the other hand, release chemicals into the environment that suppresses the growth and survival of other plants, a process known as allelopathy.

Long-term competition and short-term competition between two plant species may result in different outcomes. It is perhaps because of some critical points in the life cycles of the plants that determine the outcome of long-term competition.

As a plant interacts only with another neighboring plants, Harper (1977) referred to this "interaction" as the "neighboring effect." According to him, the neighboring effect could be negative as well as positive. The negative neighboring effects include: (1) shortage of vital resources, for example light, water, and mineral nutrients; (2) decreased environmental capacity needed for triggering the process of breaking down seed dormancy; (3) increased vulnerability to lodging; (4) increased vulnerability to pest epidemics and hazards like being grazed; (5) declined availability of pollinators; and (6) release of deleterious chemicals (allelopathy).

The negative effects emanating due to increased resource consumption by a plant at the cost of other plants could, according to Harper (1977), be best reserved as a case for competition.

The positive neighboring effects, on the other hand, include: (1) decrease in overabundant resources that could affect normal growth rate of the plants; (2) protection against pest epidemics, lodging, and grazing; and (3) increased availability of some essential nutrients, such as nitrogen in the presence of nitrogen fixers, as symbionts with neighboring plants.

Relating to the concept of sustainability, increased diversity of plants in an ecosystem adds to resilience and, thus, enhances the level of sustainability. Populations are more sustainable than a single individual in a space (ecosystem). A system with a high degree of heterogeneity is more sustainable than that with monocultures of a single species. The positive effects, however, need not counteract the negative effects (competition). Our understanding of competition among plants, thus, offers us to manage the systems, particularly the anthropogenic ecosystems, like cultivated lands, with keeping in mind the competition between various species.

COMPETITION WITHIN PLANT POPULATIONS

When there is interference of a plant with the other, it may either fail to germinate despite retaining its viability, or it may die, or it may survive and grow with limitations imposed by its environment (including the neighboring plants). Competition among plants in a population can be better understood by density-yield relations. The yield of a species in a population (monoculture) per unit area after a given period of growth increases at a lower density. However, with continuous low density, there is a linear increase in yield. At a higher density, further increase in density records a smaller progressive increase in yield. Eventually, the maximum level of yield is attained and further increase in density registers no further increase in yield. It is because the whole population (monoculture of a species) is encountered by limited environmental resources.

From the rich body of literature emerge two equations further explaining the density-yield relationship. The "reciprocal yield" relationship elucidated by Bleasdale and Nelder (1960) and Holliday (1960) was:

$$1/w = ad + b$$

where w is the mean yield per plant; d is the density; and a, b are the constants.

A general and more advanced formulation to express the density-yield relationship was evolved by Vandermeer (1984). This is:

$$w = w_m / (1 + cd^{-e})$$

where w_m is the mean yield of isolated plants; c is the measure of competition including its intensity and the area it operates within; d is the density; and e is the rate at which the competition effect decays as a function of the density between plants.

The addition of nitrogen (e.g., through nitrogenous fertilizers, like urea) or any other limiting resource might alter the optimal density of the plants as well as maximum yield, leading to a change in the values of the constants. However, this will not affect the density-yield relationships in the previously mentioned formulae.

ASYMMETRIC COMPETITION

There are two aspects of individual competing plants: dividing up and consuming the limiting resources. Larger competing individuals will obviously divide up and consume larger chunks of the limiting resources, thereby suppressing the normal growth and yield of the smaller competing individuals. Thus, asymmetric competition denotes a competing situation in which larger individuals have a disproportionate effect or obtain a disproportionate share of natural resources. The larger individuals compete in a fashion that they monopolize the limiting resources (disproportionate sharing) that happens to fulfill greater needs of relatively larger size.

INTERSPECIFIC COMPETITION: MIXED-SPECIES ANALYSIS

Heterogeneity-ridden cultures are more complex than the monocultures of a single species. A mix of species in a culture leads to increased complexity among diverse plant species. In such a case, the density of both the components in the culture influences the performance of each component.

In many cases of mixed-species cultures, it is observed that the sum total yield is more than in a case when a species is grown in a monoculture. It is because of what is referred to as complementation or complementarity, a phenomenon in which two plant species positively influence the growth of each other, which might be due to differential resource utilization by different species. Complementation depends on the nature of resource utilization by the species. If the resource requirements of the two species differ, they both would be able to utilize more of the resources available to them if they were raised together than if each of the species were grown in a monoculture. A negative neighboring effect in a monoculture is countered by a positive neighboring effect in mixed cultures. A glaring example is raising a cereal crop (family Poacea) along with a leguminous crop (family Leguminosae). Nitrogen fixed by the legume also becomes available to cereals, and such a positive influence results in higher yield in a mixed culture.

Competition in nature, in essence, is a reality. We often encounter weed eradication, an essential operation in agriculture. Weeds are the competent plants, so are the plants in the food crops—the competition between two different species. We eradicate only the competing plants that are of no, or comparatively of less, economic importance.

In natural ecosystems, all kinds of competitions within a diversity of plants occur. Despite this natural fact, enormous diversity is often encountered existing in natural ecosystems. Ecologists have inferred that species diversity enhances forest productivity (Fichtner et al. 2017) and sustainability (Singh 2018). Competition within species in a community appears to be an imperative of natural evolution. Co-existence in nature, despite competition, is a universal law, and it appears to be an indispensable dimension of an evolutionary phenomenon for continuous enrichment of communities and stabilization of natural ecosystems. This is the natural path toward the evolution of ecologically vibrant and sustainable climax communities in the biosphere.

ALLELOPATHY

Allelopathy (Greek: *Allelon* = of each other, *Pathos* = to suffer), a term coined by Hans Molisch (1856–1937) in 1937, is a biological phenomenon in which some plants produce specific biochemicals that exert an influence on germination, growth, nutrient uptake, survival, symbiotic relationship, reproduction, and yield of other plants (or microorganisms). The biochemicals released into the environment are called allelochemicals. The allelochemicals released by one organism may be beneficial or detrimental for the other organism or the whole community.

Allelopathy, in essence, is the phenomenon of chemical inhibition (or amelioration) of a species by another species. The allelochemicals can be released from any part of a plant—roots, stems, leaves, flowers, fruits, etc.—and are also present in the surrounding environment, especially soil. Not all plant species are allelopathic in nature. And yet there are some species releasing chemicals that are beneficial (e.g., growth promoting) rather than detrimental (e.g., growth inhibiting) for other species. The beneficial type of allelopathy may be referred to as positive allelopathy and the detrimental one as negative allelopathy.

Both abiotic (such as temperature and pH) and biotic factors affect the production of allelochemicals by plants. Most of the allelopathic plants have leaf stores of allelochemicals. With leaf fall, the allelochemicals get released into the soil environment and reveal their effect on the plants. Some of the allelochemicals associated with allelopathic species are given in Figure 6.1.

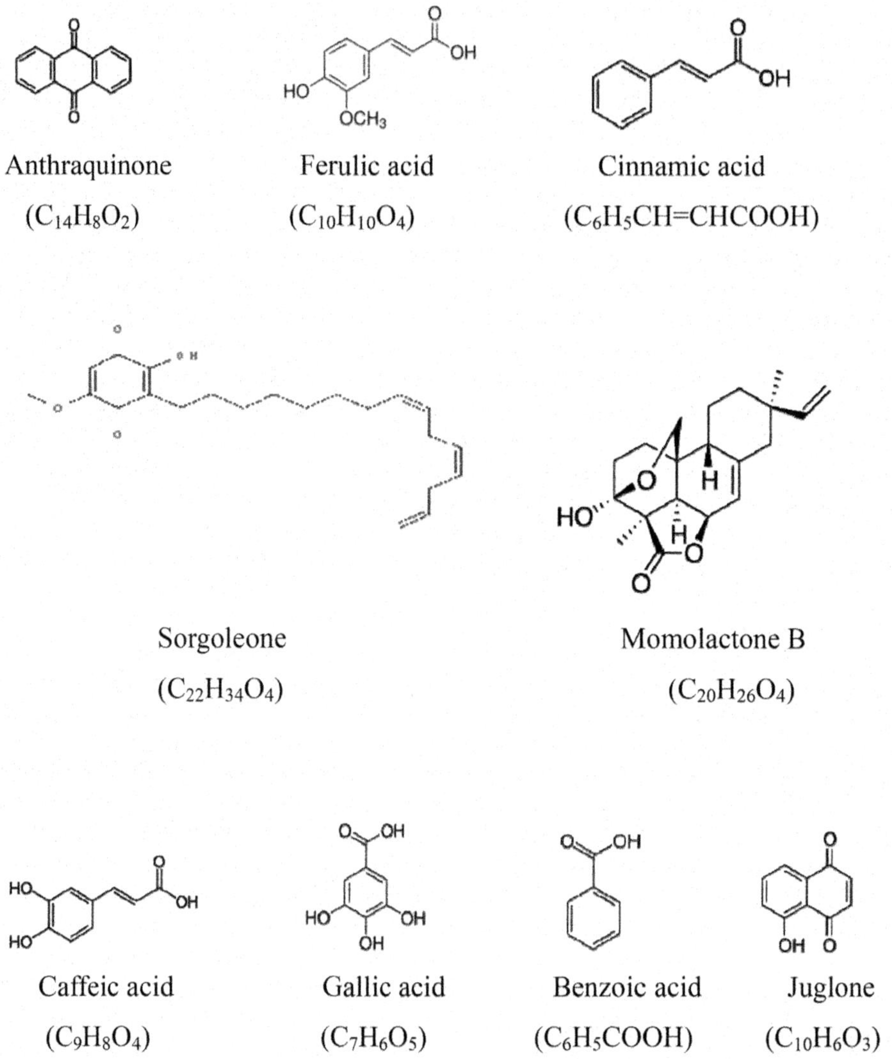

Anthraquinone
($C_{14}H_8O_2$)

Ferulic acid
($C_{10}H_{10}O_4$)

Cinnamic acid
($C_6H_5CH{=}CHCOOH$)

Sorgoleone
($C_{22}H_{34}O_4$)

Momolactone B
($C_{20}H_{26}O_4$)

Caffeic acid
($C_9H_8O_4$)

Gallic acid
($C_7H_6O_5$)

Benzoic acid
(C_6H_5COOH)

Juglone
($C_{10}H_6O_3$)

FIGURE 6.1 Structure of some of the allelochemicals produced by allelopathic plants.

CHEMICAL INTERACTIONS AMONG PLANTS

In a biocommunity, each plant interacts with others positively, negatively, or neutrally. Chemical-mediated interaction among plants, that is, allelopathy, is vital in structuring a biocommunity. The need of research and development intervention in allelopathy in forestry, cropping, and natural ecosystems is a pressing need of our contemporary world (Waller 2005).

Allelopathic plants release harmful chemicals (allelochemicals) from their roots (and also from other parts), which target other neighboring plants. Some of the allelochemicals reduce chlorophyll contents in a plant and consequently slow down or stop photosynthesis of the target plant, eventually resulting in suppressed growth or death of the plant.

Some of the allelopathic plants release allelochemicals in gaseous form. These allelochemicals are released from the small leaf pores. The allelochemical gases target the neighboring plants, suppress their growth, and/or kill them.

Allelopathic plants shed their leaves containing allelochemicals during fall season. The leaves decompose and release their noxious chemicals, which suppress the growth of their target plants.

NATURE OF ALLELOPATHY

Reduced seed germination and seedling growth are the commonly known effects of allelopathy. The nature of allelochemicals is distinguishable from synthetic herbicides. All the allelochemicals do not have a common physiological target or a common mode of action. The synthetic herbicides, on the other hand, generally have a common physiological target and mode of action. The common sites of allelochemical action are photosynthesis, specific enzyme action, nutrient uptake, pollen germination, and cell division. An allelochemical, 3-(3′,4′-dihydroxyphenyl)-l-alanine (I-DOPA) found in velvet bean has its action on amino-acid metabolism and iron-concentration equilibrium.

Different classes of chemicals—flavonoids, steroids, terpenoids, phenolic compounds, alkaloids, etc.—are involved in complex processes of allelopathic inhibition of plants. A mixture of different allelochemicals is often more deleterious than an individual chemical. Pests and diseases, herbicides, solar radiation, temperature, suboptimal nutrients, moisture, and physiological and environmental stresses also influence allelopathic reactions to a great extent. Allelochemicals can be secreted from any part of an allelopathic plant. They can also persist in soil leachates for a long time and may affect target plants at an appropriate time. The rate of biodegradation of allelochemicals may be higher than that of traditional herbicides. Non-target species may also be affected by certain allelochemicals.

One glaring example of selective allelochemical is mimosine, a free amino acid. Found in the leaves of *Leucaena leucocephala*, this non-protein amino acid inhibits the growth of other plant species but not of its own species. This tree species, of late, had been quite popular in India and was extensively planted for water and soil conservation, livestock feeding, and wasteland management and has been found detrimental for the growth of other species. Many invasive or exotic species expressing very high fecundity have strong ecological holds due to their allelopathic properties.

The ecological consequences of allelopathy, as elaborated by Koocheki et al. (2013), underline the specific nature of specific plant species governing ecological processes influencing changes in ecosystem structure and functioning. Allelochemicals released by one plant species to play a role against other plant species have long-term consequences in the evolution of natural ecosystems. These chemicals play important roles to influence and determine ecological processes, such as plant diversity, nutrient dynamics, soil fertility, microbial populations, and succession.

ALLELOPATHIC PLANTS

Not all the plants on Earth are allelopathic in nature. However, quite a large number of allelopathic plants, both natural or wild and cultivated ones, are there in nature. A list of some of such prominent allelopathic plants is given in Table 6.1. Some of the most talked-about allelopathic plants are trees such as black walnut (*Juglans nigra*), fragrant sumac (*Rhus aromaticus*), and ailanthus (*Ailanthus altissima*) and cultivated plants such as sorghum (*Sorghum bicolor*), rice (*Oryza sativa*), and pea (*Pisum sativum*). Black walnut with a respiratory-inhibitor allelochemical known as juglone found in its roots, leaves, buds, and nut hulls is one of the well-known and expert allelopathic trees.

TABLE 6.1
Allelopathic Plants and Their Allelopathic Impacts

Allelopathic Plant	Impact
Black walnut (*Juglans nigra*)	Allelochemical compound juglone reduces corn yield
Broccoli	Residue interferes with other crops of Cruciferae family
Buckwheat	Cover crop residue reduces weed menace in fava bean crop
Chaste tree	Retardation of pangola grass and pasture grass growth, but stimulation of bluestem grass growth
Chicory	Inhibition of *Amaranthus retroflexus* and *Echinochloa crusgalli*
Crabgrass	Inhibition of corn and sunflower
Desert horsepurslane (*Trienthema portulacastrum*)	Growth promotion in *Amaranthus viridis*
Eucalyptus	Suppression of wheat crop
Fescue	Weed suppression when cultivated as cover crop or residues retained as such in the field
Forage radish	Mulching of forage radish suppresses weeds in the following season
Garlic mustard	Inhibition in arbuscular mycorrhizal fungi in maple
Green spurge	Inhibition of chickpea growth
Jatropha curcas	Inhibition of tobacco and corn by extracts of leaves and roots; inhibition of germination of *Glycine max*, *Brassica juncea* and *Oryza sativa*.
Jerusalem artichoke	Residual effects on weed species
Jungle rice	Inhibition of rice crop
Lantana	Reduction in germination and growth of milkweed
Leucaena leucocephala	Reduction in wheat and turmeric yields, but increase in rice and corn yields
Mango	Inhibition of sprouting of purple nutsedge tubers
Parthenium hysterophorus	Inhibition of cereal crops
Rabbitfoot grass	Inhibition of wheat by leaf extract and mulch
Red cedar	Inhibition of lettuce-seed germination
Red maple	Inhibition of lettuce-seed germination
Rhazya stricta	Growth inhibition in corn
Rye	Weed suppression when cultivated as cover crop or residues retained as mulch in the field
Silver wattle	Inhibition of native understory species
Sour orange	Inhibition of seed germination and root growth of bermudagrass, lambsquarters, and pigweed
Sun hemp	Inhibition of vegetable-seed germination, inhibition of smooth pigweed growth
Sunflower	Cover crop residue reduces weed menace in fava bean crop
Swallow-worts	Inhibits several weed species
Swamp chestnut	Inhibition of lettuce-seed germination
Sweet bay	Inhibition of lettuce-seed germination
Teak wood	Leaf extracts inhibit jungle rice and sedge
Tifton burclover	Inhibition of wheat growth
Tree of heaven	Ailanthone, a chemical compound isolated from the tree, shows herbicidal activity like that of glyphosate and paraquat
Wheat	Weed suppression when cultivated as cover crop or residues retained as mulch in the field
Xanthium strumarium	Growth inhibition in mungbean

Source: http://edis.ifas.ufl.edu/hs186 and Rastogi, A. et al., *J. Emerg. Technol. Innov. Res.*, 6, 866–874, 2019.

ALLELOPATHY APPLICATIONS IN AGRICULTURE

Allelopathy is one of the natural phenomena used in agriculture since ancient times (Zeng 2014). While allelopathy can also stimulate plant growth, it is generally viewed as a negative attribute of a plant species for other target-plant species. Allelochemicals can help develop crops with low phytotoxic-residue contents in soil and water. This is how allelochemicals can help in wastewater treatment and recycling (Zeng et al. 2008). Many of the allelochemicals produced by allelopathic plants serve as appropriate substitutes for synthetic herbicides. Allelochemicals are ecologically friendly and environmentally safe because they are biodegradable and do not leave residues or persistent or long-term toxic effects. On the basis of specific allelopathic traits, many crop patterns can be developed (Rastogi et al. 2019), and a synergy among different crops can be created for obtaining more yields and more nutrients per unit area. Ecologically sound control of weeds, plant pathogens, soil-nitrogen conservation in croplands, etc. can be effectively managed by incorporating selective allelopathic plants (crops) in agricultural systems.

CROPPING SYSTEMS' MANAGEMENT

Allelopathy arouses immense interest due to its potential applications in agriculture. Declined crop yields due to residues from the previous crop might be on account of allelopathy. Allelopathic plants chemically compete out the plants that might not be desirable in agriculture. This is a positive attribute of allelopathy in agriculture that we can manage weeds in the crop fields in the current season as well as in the following season. The specific ability of some crops to suppress weeds, in this way, is found out by crop allelopathy (Cheng and Cheng 2015) and its applications in the management of cropping systems are pertinent and appreciable. *Jatropha curcas*-based intercropping involving *Glycine max*, *Brassica juncea*, and *Oryza sativa* are badly affected by allelopathic effects of the base crop *Jatropha*, as found in the Tarai region of Central Himalayas in India (Rastogi et al. 2019). Crop allelopathy can also be utilized to alleviate allelopathic autotoxicity and reduce the inhibitory effect among allelopathic crops. It serves to ameliorate land-utilization measures along with an increase in seasonal soil output by designing suitable intercropping systems and crop rotations. Some of the research outcomes, as documented by Cheng and Cheng (2015), suggest that:

- Nematode control in crop production can be effectively managed using bush fallow with *Chromolaena odorata* L. (an example from southwestern Nigeria).
- Weeds in the cropping systems can be dealt with in an environment friendly manner by means of crop rotation and intercropping systems.
- Intercropping involving sorghum, sesame, and soybean in cotton fields helps manage purple *Cyperus rotundus* L. (nutsedge) compared to cotton as a sole crop.
- Relay intercropping of eggplant/garlic is found beneficial in terms of higher yields and economic returns.

MULCHING FOR WEED CONTROL

Allelopathy provides an ecological solution to weed management. Instead of using extremely toxic synthetic herbicides, allelopathic applications can serve as an environment-friendly, inexpensive, and sustainable source for weed management. One of the examples of allelopathic applications is using straw mulch. It prevents environmental pollution, especially the soil pollution, and contributes to soil fertility. An environment-friendly option is the use of allelopathic plants as ground-cover species. Decomposing straw from an allelopathic crop releases allelochemicals during the process of decomposition that play a role in suppressing weeds as well as in curbing pests of various sorts prospering on a crop (Cheng and Cheng 2015).

In addition to deriving an economic advantage out of the allelopathic applications of straw mulch, we can also enrich the soil organic carbon and preserve soil moisture. Both of these soil factors enhance soil fertility and increase crop productivity. However, there is also a drawback of straw mulching: this increases the C:N ratio of the soil. Based on a research experiment cited by Cheng and Cheng (2015), it has been revealed that the straw of green wheat (*Triticum aestivum* L.) inhibits *Ipomea* weed proliferation in *Zea mays* L. and *Glycine max* L. This straw application helps reduce the need of herbicide sprays and thereby avert soil, water, and air pollution.

ECOLOGICALLY HEALTHY AGROCHEMICALS AND MICROBIAL PESTICIDES

Applications of allelopathy can be of critical importance in plant protection against weeds and herbivory. This can be enacted by applying technology worth modifying the allelochemicals for the production of plant-growth regulators and environment-friendly pesticides, which can be used in effective management of agricultural systems and in preventing pollution by avoiding conventional applications of synthetic pesticides. High degradability of allelochemicals also ensures the maintenance of soil's ecological health.

One of the examples of modifying allelochemicals involves sorgoleone, a hydrophobic chemical found in the sorghum-root exudates. This allelochemical as a wettable powder acts as a strong weed suppresser. In this modified form, sorghum plants were tolerant to their own allelochemical modified into a wettable powder. The other advantage noted was that some soil microorganisms are capable of utilizing sorgoleone as a carbon source upon its degradation in the soil into CO_2. Effective action of the formulated sorgoleone is instrumental in weed management in an environment-friendly, more effective, and natural way (Cheng and Cheng 2015).

A number of organic herbicides or plant-growth inhibitors recommended for use in organic farming cultures have been manufactured from selected allelopathic plants (Cheng and Cheng 2015). For example, a mixture of allelochemicals extracted from *Pinus* spp., Japanese cedar (*Cryptomeria japonica* D. Don), hinoki (*Chamaecyparis obtusa* Endl.), and bamboo vinegar was used for preparing an eco-friendly and effective organic herbicide for use in rice fields (Cheng and Cheng 2015). This is how allelopathy can be practically and safely used in agriculture, gardening, and forestry.

Certain allelopathic bacteria can also be of potential value in promoting plant growth and/or suppressing allelopathic weeds. The plant-growth promoting rhizobacteria (PGPR) include a variety of bacteria exerting positive influence on plants. A positive effect, like this, includes induced systemic resistance, helping reduce crop vulnerability to pathogens. Microbial pesticides accrued from certain allelopathic bacteria exhibiting PGPR attributes and reducing inhibitory effects on plants susceptible to allelopathic weeds are in great demand in organic farming cultures.

BREEDING THE ALLELOPATHIC CULTIVARS

The most promising allelopathic applications for weed management and minimizing refractory chemicals' introduction lie in the breeding of allelopathic plants in crop fields. Breeding methods generally in practice may be applicable for breeding of the selected allelopathic plants. The intended cultivars should exhibit desirable properties, such as effective weed inhibition, disease resistance, early maturity, high yields, and quality traits.

A rice cultivar named Rondo carrying the previously mentioned properties and whose weed-inhibition ability is considered superior than many commercial cultivars was raised in an organic rice-production system in Texas, USA. Spring wheat was bred using conventional methods for improved allelopathic potential. This was a product of the cross between Swedish and Tunisian cultivars with low and high allelopathic activity, respectively. The newly bred cultivar had 19% decline in weed biomass. However, 9% reduction in grain yields was also registered. Huagan 3 is regarded as the first acceptable weed-inhibiting cultivar in China (Cheng and Cheng 2015).

REDUCING NO$_3$ LEACHING AND N$_2$O POLLUTION

Leaching of the nitrates in the soil into the groundwater and surface water bodies is one of the serious pollution problems. Nitrates are responsible for methemoglobinemia, popularly known as blue baby syndrome. Nitrification of nitrogenous fertilizers, like urea, or mineralization of organic nitrogen, serves as a process for nitrogen-enrichment of the soil. An increment in soil nitrification can occur as a result of biological nitrification inhibition (BNI) in which allelopathic plants play a crucial role. There are certain nitrification-inhibiting substances (NIS) produced by plants that can be used for managing soil nitrification. There are many types of allelochemicals produced by some allelopathic plants that function as biological nitrification-inhibiting substances (BNIS). BNIS inhibits the processes of soil nitrification.

The allelochemicals, viz., hydroxamic acid, *p*-hydroxybenzoic acid, and ferulic acid excreted from wheat (*Triticum aestivum* L.) act upon soil microorganisms and inhibit soil nitrification and, thus, contribute to enhancing nitrogen uptake by plants. In addition, the allelochemicals also curb N$_2$O emissions and, in this way, help reduce environmental pollution by a greenhouse gas. Allelochemicals *Plantago lanceolata* L. oozed out into soil exerts an inhibitory effect on organic nitrogen mineralization. This suggests that this plant helps reduce soil-nitrogen leaching.

ALLELOPATHIC MECHANISMS

EFFECT ON SOIL ECOLOGY

The presence of allelopathic species and allelochemical applications exert both negative and positive influences on soil ecology (by influencing soil microorganisms) and on vegetation/crop performance, for example, growth and reproduction. In plant-to-plant interactions, the indirect allelopathic effects are more powerful than the direct ones of the selected allelochemicals (Zeng 2014). A negative feedback of soil sickness and plant growth is generated as a result of the behavior of microorganisms influenced by specific allelochemicals. Soil microorganisms inhabiting the rhizosphere, on the other hand, generate positive feedback favoring plants' allelopathic potentials. Soil bacteria can also change the very nature of an allelochemical, for example, by activating the non-toxic allelochemical and thus increasing its inhibition potential. Soil bacteria, on the other hand, can also transform a toxic allelochemical into a non-toxic one, and thus can avert, or minimize, the allelopathic effects of some weeds on the targeted plants of economic importance (see Figure 6.2).

Soil microorganisms can transform the nature of allelochemicals (e.g., non-toxic to toxic and vice versa) and, in this way, allelopathy is vital in altering soil ecology and consequent plant-to-plant interaction, eventually altering the vegetation composition in an ecosystem. PGPR, such as *Pseudomonas* that colonize plant roots, *Paenibacillus polymyxa*, endophytes, etc. are capable of altering the plant-gene expression and, in response to abiotic and biotic stress, play a crucial role in regulating allelochemical synthesis by plants and in signaling pathways to result in enhancing disease resistance and defense and the adaptation capability of the plants (Cheng and Cheng 2015), which indicates that allelopathy effectively functions at the genetic level. As the soil ecology phenomenally influences the terrestrial (above-soil) biodiversity, the allelopathy, in essence, can contribute to a decrease or substantially increase the biodiversity in an ecosystem.

EFFECT ON PHOTOSYNTHESIS

Allelochemicals released by some plants affect photosynthesis by other plant species in three ways: (1) inhibition of photosynthetic machinery, (2) damage to photosynthetic machinery, and (3) decomposition of photosynthetic pigments. A considerable decrease in the photosynthetic pigments affects photosynthesis by (1) blocking energy and electron transfer, (2) decreasing enzyme action in ATP synthesis, (3) inhibiting ATP synthesis, (4) affecting stomatal conductance and transpiration, and (5) inhibiting the overall photosynthetic process.

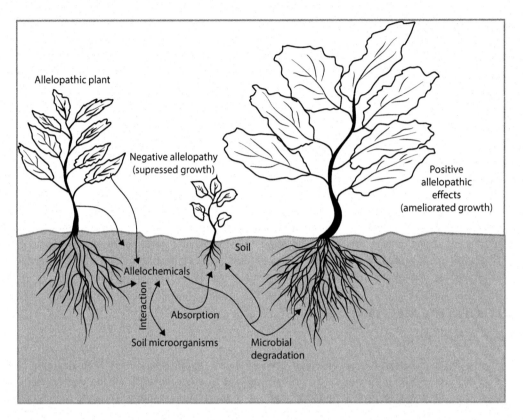

FIGURE 6.2 Soil microorganisms as modulators in interactions between allelopathic donor-receiver species. (Adapted from Cheng, F. and Cheng, Z., *Front. Plant Sci.*, 6, 1020, 2015.)

Allelochemicals, as Weir et al. (2004) found, hit photosynthesis by affecting the functions of PS II. Earlier, sorgoleone, an allelochemical released by sorghum, inhibits the photosynthetic process largely due to its influence on PS II. This allelochemical leads to the reduction of Fv/Fm of weeds and suppression of growth of the competing weeds. Ye et al. (2013) observed a decline in the active reaction centers and electron-transport chain impairment during the inhibitory effect of a dried microalga *Gracilaria tenuistipitata* on another microalga, *Phaeodactylum tricornutum*. Essential oils in the leaves of lemongrass (*Cymbopogon citrates*) were found registering a significant reduction in chlorophyll a, chlorophyll b, and carotenoids in barnyard grass (Poonpaiboonpipat et al. 2013). This suggests that essential oils in the lemongrass function as allelochemicals for the barnyard grass by affecting photosynthetic metabolism.

EFFECT ON RESPIRATION

The allelochemical impact on respiration contributes to curbing plant growth. Allelopathic plants influence respiration at the physiological and biochemical levels. The various respiration stages that allelochemicals influence are:

1. Electron transfer in mitochondria,
2. Oxidative phosphorylation,
3. ATP enzyme activity, and
4. CO_2 production.

Oxygen intake is reduced, which triggers the mechanism leading to respiration inhibition involving prevention of NADH oxidation, inhibition of ATP synthesis, decline of ATP production in mitochondria, disturbance in plant-oxidative phosphorylation, and eventually respiration inhibition. Some allelochemicals, on the contrary, can also accelerate respiration by stimulating release of CO_2.

Some notable research findings relating to the allelochemical influence on respiration as reviewed and cited by Cheng and Cheng (2015) are as follows:

- Corn pollen releases ethanol that reduces O_2 consumption and acts as electron-pathway inhibitor. Upstream of cytochrome c is the likely location of the specific inhibition site.
- Sorgoleone, an allelochemical released by sorghum, blocks electron transport at the b-c1 complex, thus interfering with the functions of mitochondria in corn and soybean seedling cells.
- Juglone, a respiratory-inhibitor allelochemical released from the roots, leaves, buds, and nut hulls of black walnut (*Juglans nigra*), disturbs root O_2 uptake in soybean and maize as this allelochemical reaches the mitochondria of the root cells.
- Monoterpenes such as alpha-pinene, camphor, and limonene influence mitochondrial respiration in the radical and hypocotyl in soybean and corn. ATP production in the mitochondria is severely inhibited by alpha-pinene. The alpha-pinene allelochemical also reduces the transmembrane potential of mitochondria, uncouples oxidative phosphorylation and electron transport inhibition, and impairs mitochondrial energy metabolism. Camphor is involved in the uncoupling of mitochondria. Limonene inhibits coupled respiration, and ATP interferes with the activity of synthetase and adenine-nucleotide-translocase complexes.

Effect on Enzyme Functions

Various allelochemicals variously influence the synthesis, functions, activities, and contents of various enzymes engaged in various biochemical and physiological mechanisms in plants. For example, caffeic acid, catechol, and chlorogenic acid inhibit the main enzyme λ-phosphorylase, which participates in the process of seed germination. Tannic acid inhibits peroxidase (POD), catalase, and cellulase enzymes. Synthesis of amylase and acid-phosphatase in the endosperm is also suppressed by tannic acid.

Chrysanthemum (*Chrysanthemum indicum* L.) releases an allelochemical in the rhizosphere. An extract of this plant in the rhizospheric soil inhibits root dehydrogenase and nitrate reductase activities (Zhou et al. 2010).

Influence on Plant-Growth Regulator System

Allelochemicals can strike an imbalance among various phytohormones responsible for plant growth or can adversely affect the plant-growth regulator system, leading to inhibition in seed germination and overall growth and development of the targeted plants. Most of the phenolic compounds, which are allelochemicals in nature, stimulate activity of indole 3 acetic acid (IAA) oxidase and inhibit a POD reaction with IAA-bound gibberellic acid (GA) to affect endogenous hormone production. Some other research findings relating to allelochemical's influence on the plant-growth regulator system are as follows:

- In cell suspension cultures of *Pyrus communis*, salicylic acid inhibited ethylene synthesis (cited by Cheng and Cheng 2015).
- Wheat-seedling growth was found suppressed due to IAA, GA_3, and cytokinin accumulation when treated with high ferulic acid, an allelochemical, with a simultaneous abscisic acid.

- With treatment by allelochemicals in rice-aqueous extract, IAA oxidase activity of a barn-yard was stimulated with a reduction in IAA levels followed by seedling-growth inhibition (cited by Cheng and Cheng 2015).
- Cyanamide in small concentrations (1.2 mM) struck an imbalance in the homeostasis of auxin and ethylene in the roots of *Solonum lycopersicum* L. (tomato), thus disturbing the growth-regulating system (Soltys et al. 2012).
- Leaf aqueous extract of *Jatropha curcas* L. in various concentrations inhibited plumule and radicle growth in two varieties each of soybean (*Glycine max*), mustard (*Brassica juncea*), and rice (*Oryza sativa*) (Rastogi et al. 2019) (Figure 6.3a–c).

Effect on Antioxidant System

There are allelochemicals that cause victim plants to produce reactive oxygen species (ROS) and induce an alteration in the activity of antioxidant enzymes, such as peroxidase, ascorbic-acid peroxidase, and superoxide dismutase to resist oxidative stress. For instance, activities of peroxidases, proteases, and polyphenol oxidases are altered by caffeic acid, an allelochemical, during the processes of root development and reduce hypocotyls phenolics in *Phaseolus aureus* (Batish et al. 2008).

Effect on Protein and Nucleic-Acid Synthesis

Allelochemicals released by certain plant species target DNA, RNA, protein synthesis, and interfere with the metabolism of these biomolecules. A few allelochemicals, especially the alkaloids, induce inhibition in DNA polymerase I and thwart DNA transcription and translation. Some allelochemicals bring an increase in the temperature of DNA cleavage. Some of them also inhibit protein synthesis in various ways by inhibiting amino-acid absorption and transport. Inhibition of protein synthesis results in decreased growth rates of the plants. Many alkaloids and phenols, ferulic acid, and cinnamic acid are involved in inhibiting protein synthesis. All the phenolic acids playing the role of allelochemicals interfere with the integrity of DNA and RNA.

Effect on Water and Nutrient Uptake

A number of allelopathic plants inhibit water utilization by the plants they target and thus push them into a state of water stress. These plants also adversely affect uptake of nutrients by their victim species. Inhibitions of Na$^+$ and K$^+$-ATPase responsible for the absorption and transport of ions at the cell-plasma membrane is also a role many allelopathic plants play against their target species, followed by suppression of the cellular absorption of Na$^+$ and K$^+$ ions. The effect of some allelochemicals on mineral-nutrient uptake by targeted plant species is presented in Table 6.2.

It is not just allelochemical but its concentration that matters for affecting a plant's mineral nutrition. In some cases, the effect of an allelochemical might be negative or positive depending on its concentration. For instance, dibutyl phthalate in high concentrations has negative implications by inhibiting the uptake of N, P, and K, the three nutrients vital for plant nutrition. The same allelochemical in low concentration, on the other hand, increases N absorption.

Effect on Cell Division and Elongation

Some allelopathic plants release monoterpenoids that carry allelochemical properties, directly hitting cell division and DNA synthesis in plant meristems. The five well-known monoterpenoids with potential allelopathic effects leading to impairment of cell division and DNA synthesis processes

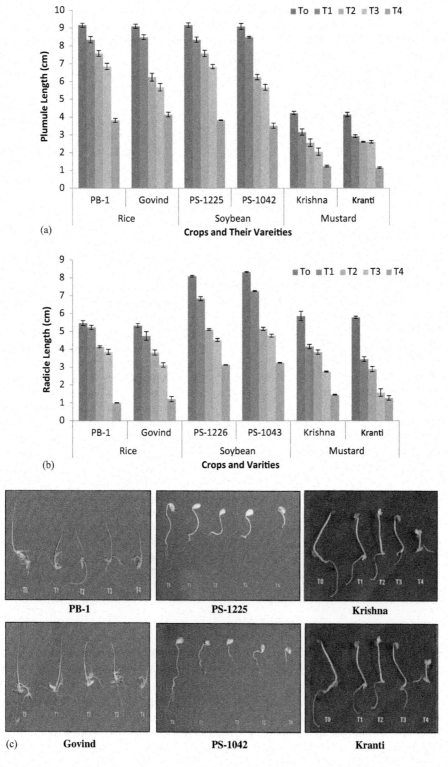

FIGURE 6.3 Allelopathic effects of *Jatrophas* L. leaf aqueous extract of varying concentrations (T0: control; T1: 5%; T2: 10%; T3: 15%; T4: 20%) on (a) plumule, (b) radicle length (cm) in rice, mustard, and soybean, and (c) radicles and plumules of the germinating seeds. (From Rastogi, A. et al., *J. Emerg. Technol. Innov. Res.*, 6, 866–874, 2019.)

TABLE 6.2

Allelochemical Interaction with Targeted Species Affecting Mineral Nutrition of the Plants

Allelochemical	Allelopathic Plant	Targeted Plant Species	Mineral Nutrient Affected	References
Ferulic acid	Rice, oat, seeds of apples and oranges	Corn seedlings	NH_4^+ and NO_3^-	Bergmark et al. (1992)
Ferulic acid	Rice, oat, seeds of apples and oranges	Wheat seedlings	NH_4-N, NO_3-N	Yuan et al. (1998)
Benzaldehyde	Apple, almond, apricot, cherry kernels	Wheat seedlings	NH_4-N, NO_3-N	Yuan et al. (1998)
4-tert-Butylbenzoic acid		Wheat seedlings	NH_4-N, NO_3-N	Yuan et al. (1998)
Cinnamic acid	Accumulated in the soil	Cucumber seedlings	Na^+, Ca^{2+}, Mg^{2+}, Fe^{2+}, NO_3^-, SO_4^{2-}	Yu and Matsui (1997)
Cinnamic acid	Cucumber root exudates	Cucumber	Dehydrogenase, ATPase, nitrate reductase	Wu et al. (2007)
p-Hydroxybenzoic acid	Cucumber root exudates	Cucumber	Dehydrogenase, ATPase, nitrate reductase, root uptake of K^+, NO_3, $H_2PO_4^-$	Wu et al. (2007)
Sorgoleone	Sorghum root exudates	Peas, soybeans, corn	Water-solute uptake, H^+-ATPase, proton function across root cell plasmalemma	Jesudas et al. (2014)
Juglone	Black walnut	Peas, soybeans, corn	H^+-ATPase, proton function across root-cell plasmalemma, solute and water uptake, etc.	Ercisli et al. (2005)
Trans-cinnamic acid	Cucumber root exudates	Maize seedlings	Net nitrate uptake, plasma-membrane H^+-ATPase activity	dos Santos et al. (2008)
Ferulic acid	Rice, oat, seeds of apples and oranges	Maize seedlings	Net nitrate uptake, plasma-membrane H^+-ATPase activity	dos Santos et al. (2008)
p-Coumaric acid	Peanuts, tomato, carrot, basil, garlic, wine, vinegar	Maize seedlings	Net nitrate uptake, plasma-membrane H^+-ATPase activity	dos Santos et al. (2008)

are: 1,8-cineole, alpha-pinene, beta pinene, camphor, and camphene. There are several examples, as reviewed by Cheng and Cheng (2015) that reveal that many allelochemicals impart an inhibitory influence on cell division and elongation (Table 6.3).

Not only does the inhibition of cell division and of cell elongation take place, but also the shape and structure of plant cells are influenced as a result of allelochemical reactions against certain species. For example, eucalyptol and camphor have been shown to widen and shorten root cells, increase vacuole numbers, and cause nuclear abnormalities (Cheng and Cheng 2015). Allelochemicals produced by *Nepeta meyeri* Benth. and *Convolvulus arvensis* L. may bring alterations in random amplification of the polymorphic DNA (RAPD) profile in the targeted species.

TABLE 6.3
Inhibitory Influence of Allelochemicals on Cell Division and Elongation Processes

Allelochemical	Targeted Species: Effect of the Allelochemical	References
2-3H-benzoxazolinone (BOA)	G2-M of lettuce: mitotic cell division suppression	Sanchez-Moreiras et al. (2008)
Sorgoleone	A potential herbicide, inhibits photosynthetic machinery and electron transport in chloroplast and mitochondria; reduction in cell numbers in each cell division period, suppresses water and solute uptake	Jesudas et al. (2014)
BOA and 2,4-dihydroxy-1,4(2H)-benzoxazin-3-one (DIBOA)	Cucumber: inhibition of regeneration and growth of root cap cells	Jonczyk et al. (2008)
Aqueous leaf extracts from *Datura stramonium* L.	In maize and sunflower: (1) negative effect on primary root elongation in maize; (2) lateral-root inhibition in maize; (3) inhibition of germination in sunflower	Pacanoski et al. (2014)
Ethyl acetate fraction of *Aglaia odorata* Lour. leaves	*Allium cepa* roots: damage of chromatin organization and mitotic spindle	Teerarak et al. (2012)

PHYTOSOCIOLOGY FOR A GREENER PLANET

Phytosociological relationships in natural and anthropogenic ecosystems determine many dimensions of ecological systems. Competitions of all types within and among plant populations, as discussed in the beginning of this chapter, are crucial in determining phytosociological relationships. However, allelochemical relationships among plant species play a more conspicuous and critical role, phenomenally influencing the ecological processes leading to significant alterations in structures and functional performance of the ecosystems. Allelochemicals released by allelopathic plants are potent enough to induce dynamic changes in ecosystems leading to even new patterns of evolution. Allelopathic plants' role in changing water and nutrient uptake and nutrient dynamics imparts an influence on soil chemistry and characteristics and microbial ecology, including mycorrhizae and populations of nitrogen-fixing bacteria. Allelopathic plants induce vulnerability in ecosystems inviting widespread onslaughts of invasive species, especially in natural ecosystems. Allelopathic compounds carry potential to significantly determine species richness, ecological succession, and thereby phytodiversity in an ecosystem. Thus, allelochemical relationships prevailing in an ecosystem are phenomenal in determining climax vegetation, ecosystem structures, and—eventually— ecosystem functioning. All these allelopathic attributes in nature—positive as well as negative—are interrelated with each other (Figure 6.4) and have been phenomenal in the evolution of natural ecosystems.

Allelopathic relationships influencing phytosociology in ecosystems can be used in accelerating ecosystem-development processes involving specific natural associations prevailing among native plant species. Positive attributes of allelopathy, incorporation of symbionts, and amelioration of microbial ecology, may be vital in suppressing invasive species playing ecocidal roles in natural ecosystems and reestablishing all the native species that have been integral components of natural ecosystems. In agricultural systems, allelopathy can be of vital use in creating symbiosis and synergy among a variety of crop plants, cropping patterns and systems, and in increasing crop productivity.

On the basis of their allelopathic relationships, we may prepare a list of all the plants with positive allelopathic traits and selectively allow them to practically regenerate forests through protection,

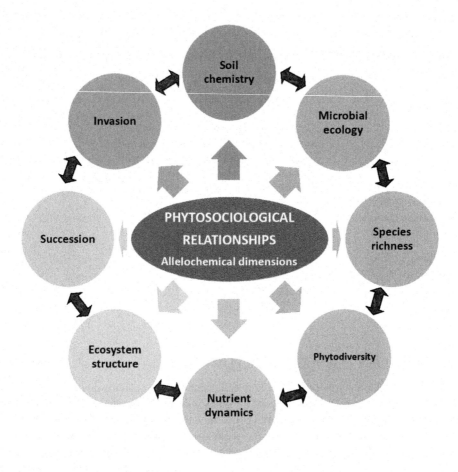

FIGURE 6.4 Phytosociological relationships as influenced by various allelopathic attributes.

conservation, and intensive plantation. In our afforestation programs, allelopathy could be used as an effective tool to boost nature's photosynthetic capacities and establish healthier, more vibrant, and ecologically sustainable ecosystems. And, thus, the prevailing allelopathic relationships among a variety of plants can be of wide applications for enhancing the processes of ecological regeneration and ecosystem restoration as necessary ingredients of the strategies helping us advance toward a climate-smart planet.

SUMMARY

All plants prevail amid a sort of competition for environmental resources, which is essential for their better survival, potential growth, and sustainable existence. Competition might be within plant populations (intraspecies as well as interspecies), and it is often negative, but could occasionally be positive. Allelopathy, a term coined by Molisch in 1937, is a biological phenomenon in which some plants produce specific biochemicals called allelochemicals that affect relationships, performance, and survival of the neighboring plants. It is a phenomenon of chemical inhibition of a species by another (allelopathic) species. Allelochemicals can be released from any part of a plant species and could also be present in the surrounding environment (soil). All species are not allelopathic in nature, and allelopathy could be positive in some cases. Some interesting allelochemicals are anthraquinone, ferulic acid, cinnamic acid, sorgoleone, momilactone B, caffeic acid, gallic acid, benzoic acid, and juglone.

All the allelochemicals do not have a common physiological target or a common mode of action. The common sites of allelochemical action are photosynthesis, specific enzyme action, nutrient uptake, pollen germination, and cell division. An allelochemical, 3-(3′,4′-dihydroxyphenyl)-l-alanine (l-DOPA) found in velvet bean has its action on amino-acid metabolism and iron-concentration equilibrium. A mixture of different allelochemicals (for example, flavonoids, steroids, terpenoids, phenolic compounds, and alkaloids) is often more deleterious than an individual chemical. Allelopathic reactions are greatly influenced by pests and diseases, herbicides, solar radiation, temperature, suboptimal nutrients, moisture, and physiological and environmental stresses.

The allelochemicals play important roles in influencing and determining ecological processes, such as plant diversity, nutrient dynamics, soil fertility, microbial populations, and succession and may have long-term consequences in the evolution of natural ecosystems. Some of the allelopathic plants that often appear in discussion are black walnut (*Juglans nigra*), fragrant sumac (*Rhus aromaticus*), and ailanthus (*Ailanthus altissima*) and cultivated plants such as sorghum (*Sorghum bicolor*), rice (*Oryza sativa*), and pea (*Pisum sativum*).

Widespread allelopathic applications in agriculture can be witnessed frequently. Allelochemicals can be useful in developing crops with low phytotoxic residue contents in soil and water. Many of the allelochemicals produced are found serving as appropriate substitutes of synthetic herbicides. Allelochemicals are ecologically friendly and environmentally safe because they are biodegradable and do not leave residues or persistent or long-term toxic effects. On the basis of specific allelopathic traits, many cropping patterns can be developed and a synergy among different crops can be created for obtaining more yields and more nutrients per unit area. Ecologically sound control of weeds, plant pathogens, soil-nitrogen conservation in croplands, etc. can be effectively managed by incorporating selective allelopathic plants (crops) in agricultural systems. Allelopathy can serve as a cornucopia of benefits in cropping-system management, mulching for weed control, obtaining ecologically healthy agrochemicals and microbial pesticides, breeding of allelopathic cultivars, reducing NO_3 leaching, curbing N_2O pollution, etc.

Allelopathic mechanisms exert their influences on soil ecology, photosynthesis, respiration, enzyme functions, the plant-growth regulator system, antioxidant system, protein and nucleic-acid synthesis, water and nutrient uptake, and cell division and elongation.

Allelochemical relationships among plant species play a conspicuously critical role, phenomenally influencing the ecological processes leading to significant alterations in phytosociological relations in an ecosystem. Allelopathic relations are potent to significantly affect soil chemistry, microbial ecology, nutrient dynamics, phytodiversity, species richness, species' invasion, ecological succession, and ecosystem structure and, thus, to determine the natural evolution of ecosystems. Allelopathic relationships can be selectively applied for enhancing the processes of ecological regeneration and ecosystem restoration.

REFERENCES

Batish, D. R., Singh, H. P., Kaur, S., Kohli, R. K. and Yadav, S. S. 2008. Caffeic acid affects early growth, and morphogenetic response of hypocotyl cuttings of mung bean (*Phaseolus aureus*). *J. Plant Physiol.* 165: 297–305. doi:10.1016/j.jplph.2007.05.003.

Bergmark, C. L., Jackson, W. A., Volk, R. J. and Blum, U. 1992. Differential inhibition by ferulic acid of nitrate and ammonium uptake in *Zea mays* L. *Plant Physiol.* 98: 639–645.

Bleasdale, J. K. A. and Nelder, J. A. 1960. Plant population and crop yield. *Nature* 188: 342.

Cheng, F. and Cheng, Z. 2015. Research progress on the use of plant allelopathy in agriculture and the physiological and ecological mechanisms of allelopathy. *Front. Plant Sci.* 6: 1020. doi:10.3389/fpls.2015.01020.

dos Santos, W. D., Ferrarese, M. L. L. and Ferrarese-Filho, O. 2008. Ferulic acid: An allelochemical troublemaker. *Funct. Plant Sci. Biotechnol.* 2 (1): 47–55.

Ercisli, S., Esitken, A., Turkkal, C. and Orhan, E. 2005. The allelopathic effects of juglone and walnut leaf extracts on yield, growth, chemical and PNE compositions of strawberry cv, fern. *Plant Soil Environ.* 51 (6): 283–287.

Fichtner, A., Hardtle, W., Li, Y., Bruelheide, H. and Kunz, M. 2017. From competition to facilitation: How tree species respond to neighbourhood diversity. *Ecol. Lett.* 20 (7): 892–900.

Harper, J. L. 1977. *Population Biology of Plants*. London, UK: Academic Press.

Holliday, R. 1960. Plant population and crop yield. *Nature (London)* 186: 22–24.

Jesudas, A., Kingsley, J. and Ignacimuthu, S. 2014. Sorgoleone from Sorghum bicolor as a potent bioherbicide. *Res. J. Recent Sci.* 3: 32–36.

Jonczyk, R., Schmidt, H., Osterrieder, A., Fiesselmann, A., Schullehner, K., Haslbeck, M., Sicker, D. et al. 2008. Elucidation of the final reactions of DIMBOA-glucoside biosynthesis in maize: Characterization of B*x*6 and B*x*7. *Plant Physiol.* 146: 1153–1163. doi:10.1104/pp.107.111237.

Koocheki, A., Lalegani, B. and Hosseini, S. A. 2013. Ecological consequences of allelopathy. In: Cheema, Z., Farooq, M. and Wahid, A. (eds.) *Allelopathy*. Berlin, Germany: Springer, pp. 23–38.

Pacanoski, Z., Velkoska, V., Tyr, S. and Veres, T. 2014. Allelopathic potential of jimsonweed (*Datura stramonium* L.) on the early growth of maize (*Zea mays* L.) and sunflower (*Helianthus annuus* L.). *J. Cent. Eur. Agric.* 15 (3): 198–208. doi:10.5513/JCEA01/15.3.1474.

Poonpaiboonpipat, T., Pangnakorn, U., Suvunnamek, U., Teerarak, M., Charoenvying, P. and Laosinwattana, C. 2013. Phytotoxic effects of essential oil from *Cymbopogon citrates* and its physiological mechanisms on barnyard grass (*Echinochloa crus-galli*). *Ind. Crops. Prod.* 41: 403–407.

Rastogi, A., Singh, V. and Arunachalam, A. 2019. Allelopathic effects of aqueous leaf extract of *Jatropha curcas* L. on food crops in the Himalayan foothills. *J. Emerg. Technol. Innov. Res.* 6 (6): 866–874.

Sanchez-Moreiras, A. M., De La Pena, T. C., Reigosa, J. M. 2008. The natural compound benzoxazolin-2(3H)-one selectively retards cell cycle in lettuce root meristems. *Phytochemistry* 69: 7172–2179. doi:10.1016/j.phytochem.2008.05.014.

Singh, V. 2018. *An Analysis of Mountain Agro-Ecosystems*. Beau Bassin, Mauritius: Lambert Academic Publishing. 72 p.

Soltys, D., Rudzinska-Langwald, A., Wisniewska, A. and Bogatek, R. 2012. Inhibition of tomato (*Solanum lycopersicum* L.) root growth by cyanamide is due to altered cell division, phytohormone balance and expansion gene expression. *Planta* 236: 1629–1638.

Teerarak, M., Charoenying, P. and Laosinwattana, C. 2012. Physiological and cellular mechanisms of natural herbicide resource from *Aglaia odorata* Lour. on bioassay plants. *Acta Physiol. Plant.* 34: 1277–1285.

Vandermeer, J. 1984. Plant competition and the yield density relationship. *J. Theor. Bio.* 109: 393–399.

Waller, G. R. 2005. Introduction: Reality and future of allelopathy. In: Masias, F. A., Galindo, J. C. G., Molinillo, J. M. G. and Cutler, H. G. (eds.) *Allelopathy: Chemistry and Mode of Action of Allelochemicals*. Boca Raton, FL: CRC Press, pp. 1–12.

Weiner, J. 1994. Competition among plants. *Treballs de la SCB* 44: 99–109.

Weir, T. L., Park, S. W. and Vivanco, J. M. 2004. Biochemical and physiological mechanisms mediated by allelochemicals. *Curr. Opin. Plant Bio.* 7: 472–479.

Wu, F.-Z., Wang, X.-Z., Wang, S.-M. and Zhou, X. 2007. Effects of cinnamic acid application on the contents in cucumber plants and soil. *Allelopathy J.* 20(2): 363–370.

Ye, C., Zhang, M. and Yang, Y. 2013. Photosynthetic inhibition on the microalga Phaeodactylum tricornutum by the dried macroalga *Gracilaria tenuistipitata*. In: Tang, X. F., Wu, Y., Yao, Y. and Zhang, Z. Z. (eds.) *Energy and Environment Materials*. Beijing: China Academic Journal Electronic Publishing House, pp. 725–731.

Yu, J. Q. and Matsui, Y. 1997. Effects of root exudates of cucumber (*Cucumis sativus*) and allelochemicals on ion uptake by cucumber seedlings. *J. Chem. Ecol.* 23: 817–827.

Yuan, G. L., Ma R. X., Liu, X. F. and Sun, S. S. 1998. Effect of allelochemicals on nitrogen absorption of wheat seedling. *Chin. J. Ecol. Agric.* 3: 9–41.

Zeng, R. S. 2014. Allelopathy-the solution is indirect. *J. Chem. Ecol.* 40: 515–516.

Zeng, R. S., Mallik, A. U. and Luo, S. M. 2008. *Allelopathy in Sustainable Agriculture and Forestry*. New York: Springer Press.

Zhou, K., Wang, Z. F., Hao, F. G. and Guo, W. M. 2010. Effects of aquatic extracts from different parts and rhizospheric soil of *Chrysanthemum* on the rooting of stem cuttings of the same species. *Acta Bot. Bor. Occid. Sin.* 76: 2–768.

WEBSITES

http://edis.ifas.ufl.edu/hs186

http://www.ncbi.nim.nih.gov/pmc/articles/PMC4647110/

https://www.maximumyield.com/definition/777/allelopathy

7 High-Altitude Physiology

Mountains seem to answer an increasing imaginative need.

Robert MacFarlance in *Mountains of the Mind*

ALTITUDE TYPES FROM SEA LEVEL

Geographical regions of the earth that are considerably higher from mean sea level (MSL) are referred to as high altitudes. All the mountains, hills, and uplands of the world constitute about 20% of the geographical areas of the land on Earth and are high-altitude regions. Altitude criterion will also apply in the atmosphere during aviation, where the five altitude types considered are true altitude (elevation above sea level), indicated altitude (altitude according to altimeter), density altitude (the density of the air according to altitude in the International Standard Atmosphere), pressure altitude (the air pressure in terms of altitude in the International Standard Atmosphere), and absolute altitude (altitude in accordance with the distance above the ground just below). The physiological effects at these altitudes in the atmosphere would accrue to human beings traveling by planes or jet planes. However, the environmental factors in the atmosphere would vary from those operating on Earth's surface. Since the plants naturally thrive in Earth's diverse geographical regions, the atmospheric altitude would hardly be a criterion for them from the viewpoint of their physiological studies.

Although the notion of "high altitude" is considered to begin at 2,400 m from the MSL, lower altitudes than this also affect plant communities, often phenomenally. Although altitude is defined based on the context of the study we carry out (geographical survey, aviation, geometry, atmospheric pressure, etc.), but here we shall focus on the altitudes in the context of the high-altitude regions where natural plant communities prevail and/or where crop cultivation takes place. Hills and mountains of the world are such regions where vegetation types exert a significant influence on global climate patterns.

Undulated lands, uplands, hills, and mountains are altogether distinguishable from the plane or even lands. Let us hereafter use the term "mountains" to refer to the high-altitude lands with respect to operating environmental factors and physiological influences of the highland habitats. Whereas the plane lands have two-dimensional spatiality, the mountain habitats are characterized by three-dimensional spatiality. Because of this additional spatiality dimension and associated conditions of altitude and slope aspect, mountain habitats experience a different set of environmental factors that exert a characteristic physiological behavior on the plant kingdom blossoming in mountain habitats.

Because altitude is the major factor determining phytosociological patterns in mountain regions, habitats are often categorized on the basis of this criterion. There are, in this way, five altitude types in the mountains, viz.:

Low altitude: below 1,200 MSL
Mid-altitude: 1,200–1,800 MSL
High altitude: 1,800–2,400 MSL
Very high altitude: 2,400–3,000 MSL
Alpine zone: 3,000–3,500 MSL
Glaciers: above 3,500 MSL

This altitudewise land classification is often used in the case of the Himalayan Mountains reference list (Singh 1998, 2018a). Zonation on an altitude basis, however, is not very strict. Various workers have classified mountain regions on the basis of different altitude ranges, according to the types of the studies executed by them. Classification based on altitude of the highland habitats, in fact, would vary from region to region and from one area to the other in the same region. Altitudes have also been delineated in accordance with the objectives of the studies, for example, socioeconomic studies, agricultural studies, and livelihood studies.

HIGH-ALTITUDE SPECIFICITIES

The key characteristics distinguishing high-altitude mountain habitats from other habitats on Earth are referred to as mountain specificities. These have been extensively discussed by Jodha (1990, 1993). Let us attempt to explain these in the context of high-altitude characteristics, which are almost indistinguishable with mountain specificities and will be looked at with a broader context of the operating factors, both environmental and anthropogenic, that are vital for influencing plant physiology and, hence, for plants' survival, reproduction, and sustenance in high-altitude habitats.

There are, in this context, six high-altitude specificities, viz. diverse ecological niches, high degree of biodiversity, fragility, adaptation mechanisms, marginality, and limited accessibility. Among these specificities, the first three are the inherent, natural characteristics of the high-altitude habitats while the latter three arise out of anthropogenic factors but are vital to influence high-altitude ecology and, therefore, all phytosociological aspects, including plant physiology.

DIVERSE ECOLOGICAL NICHES

Owing to their specific physical and environmental features driven by altitudes, slope orientation, and proximity to glaciers (as in case of the Himalayas and the Alps), there exists a diversity of ecological niches specific for specific plant species and their genotypes. Vegetation types change according to altitudes. Mountain valleys harbor the species and are often more appropriate for crop cultivation. The valleys are often flatlands amidst mountain habitats generally exploited for agricultural production. Lower and mid-altitudes of the mountains provide an appropriate environment and climatic conditions for a set of different plant communities. Alpine habitats are yet the unique plant communities on high altitudes, between timberline and glaciers.

The niches not only change according to altitude but also vary from each other even at short distances on the same altitude. For example, there is an enormous niche difference according to slope aspects, which are of importance in creating a microclimate variation. A south-oriented slope at a particular altitude is relatively warmer and drier than the north-oriented slope; southeast is the warmest and the northwest is the coolest at the same altitude. The vegetation types also vary according to these microclimatic conditions that become the basis of niche diversity. Edaphic factors also determine the types of the vegetation, and hence also contribute to the niche diversity.

It is thanks to a great variety of ecological niches in the high-altitude habitats that mountain people derive great economic advantage by managing cultivation practices in tune with the specificity of ecological niches. Mountain farmers in the Indian Himalayas, for example, exploit nine farming situations for cropping, horticulture, forestry, floriculture, animal husbandry, and all other land-based activities. These are (Singh 2018a):

1. Valley irrigated
2. Valley rain fed
3. Lower hills north slope
4. Lower hills south slope
5. Mid-hills north slope
6. Mid-hills south slope

7. High mountains north slope
8. High mountains south slope
9. Alpine pastures

A niche with a specific environmental setup is conducive to cultivating specific crops (cereals, coarse cereals, pseudo-cereals, oil-seed crops, vegetable crops, fruits, medicinal and aromatic plants, etc.). The niche-based cultivation practices in the high-altitude ecosystems further add to the biodiversity in the region.

BIODIVERSITY

In mountain areas, unlike in the plains, immense variation prevails among and within eco-zones, even within short distances. Immense diversity in the mountains is an attribute of the interactions among a number of environmental factors, for example, altitude, elevation, slope orientation and steepness, moisture regime, edaphic factors, geologic conditions, winds, precipitation, microclimatic conditions, mountain mass, and relief of terrain. Traditional cultural patterns constituting the core of local livelihoods further add to the biological heterogeneity in mountain areas. Diversity of mountain environmental conditions creates several distinct microenvironments and further influence evolution of biological diversity and native genetic resources (Jodha 1993; Partap 1993; Singh 2018a).

There is evidence that some mountain plants have evolved in response to their particular altitudinal environmental factors. High-altitude plants, in comparison to their lowland counterparts, have an increased ability to fix CO_2 from lower than sea-level concentrations. In addition, these plants use high-irradiance levels more efficiently and photosynthesize and grow at considerably lower temperatures (Friend and Woodward 1990; Singh 2018a).

In India, there are eight subcenters that are exceptionally rich in biodiversity. The three of these, namely the Western Himalayas, the Eastern Himalayas, and the North Eastern Himalayas, fall in the mountainous region. The other subcenters include the Gangetic Plains, the Indus Plains, the Western Ghats, the Eastern Ghats, and the Andaman and Nicobar Islands (Singh 2018). Jodha (1993) and Partap (1993) refer to mountain biodiversity as "extreme." Both the biodiversity hot spots that are there in India lie in the mountainous region, viz. the Western Ghats and the Eastern Himalayas.

The Himalayan mountains have especially been the source of a spectacular diversity of several fruits, tuberous vegetables, sugar-yielding plants, cereals, pulses, oil-yielding plants, spices, and their close relatives that number 155 (Arora and Nayar 1984). Wild relatives of today's economic plants found in the Himalayan region are presented in Table 7.1.

"Most mountain ecosystems host more biodiversity than any other part of the world—more even than lowland rain forests," said Koffi Annan, the erstwhile UN Secretary General in his message to the Global Mountain Summit organized on the occasion of International Mountain Year 2002. "Mountain landscapes harbor much of the world's remaining heritage, including many species of plants and animals found nowhere else, as well as the original varieties of many of our major crops."

FRAGILITY

Verticality (against gravitation) of high-altitude ecosystems measured in altitudes, coupled with slope steepness (against gravitation), impart a dimension of fragility to the mountain ecosystems of the world. The associated factors exacerbating fragility include geologic, edaphic, and biotic (especially anthropogenic) factors. Fragility reduces high-altitude ecosystems' capacities to withstand even small degrees of disturbance. Himalayan mountains being the world's youngest and highest geological features are ecologically sensitive and prone to extensive physical damage due to a very high degree of inherent fragility.

TABLE 7.1

Wild Relatives of Economic Plants in the Indian Himalayan Region

A. Northwest and Northern Himalayas

Fruits	*Elaeagnus hortensis, Ficus palmate, Fragaria indica, Morus* spp., *Prunus acuminata, P. cerasoides, P. cornuta, P. nepalensis, P. prostrata, P. tomentasa, Pyrus buccata, P. communis, P. kumaoni, P. pashia, Ribes glaciale, R. nigrum, Rubus ellipticus, R. fruticosus, R. lanatus, R. lasiocarpus, R. molluccanus, Zizyphus vulgaris.*
Vegetables	*Abelmoschus manihot* (tetraphyllus forms), *Cucumis hardwickii, C. callous, Luffa echinata, L. graveolens, Solanum incanum, S. indicum, Trichosanthes multiloba, T. himalensis.*
Cereals and Millets	*Aegilops tauschii, Avena barbeata, A. fatua* var. *fatua, A. ludoviciana, Digitaria sanguinalis, Elymus dalutricus, E. dasystachys, E. nutans, E. distans, E. orientale, Hordeum glaucum, H. spontaneum, H. turkestanicum, Penisetum orientale.*
Legumes	*Cicer microphyllum, Lathyrus aphaca, Moghania vestita, mucuna capitata, Trigonella emodi, Vigna capensis, V. radiata* var. *sublobata, V. umbellata.*
Oilseeds	*Lapidium capitatum, L. latifolium, L. draba, L. Ruderale.*
Fibers	*Linum perenne.*
Spices and Condiments	*Allium rubellum, A. schoenoprasum, A. tuberosum* and other species, *Carum bulbocastinum.*
Miscellaneous	*Saccharum filifolium, Miscanthus nepalensis.*

B. Eastern and Northeastern Himalayas

Fruits	*Fragaria indica, Morus* spp., *Myrica esculenta, Prunus acuminata, P. cornuta, P. jenkinsii, P. nepalensis, Pyrus pashia, Ribes glaciale, Rubus lineatus, R. ellipticus, R. lasiocarpus, R. molluccanus, R. reticulates, Citrus assamensis, C. ichangensis, C. indica, D. hookeriana, Eriobotya angustifolia, Mangifera sylvatica, Musa balbisiana complex, M. manii, M. nagensium, M. sikkimensis, M. supreba, M. velutina.*
Vegetables	*Abelmoschus manihot* (pungens forms), *Luffa graveolens, Neoluffa sikkimensis, Cucumis hystrix, C. callosus, Momordica dioica, M. cochinchinesis, M. macrophyllata, M. subangulata, Trichosanthes cucumerina, T. dioica, T. dioaelosperma, T. khasiana, T. ovata, T. trumcata, Solanum indicum and tubertypes, Allocasia macrorhiza, Amorphophallus bulbifer, Colocasia esculenta, Dioscorea alata, moghania vestica, Vigna capensis.*
Cereals and Millets	*Digitaria cruciata, Hordeum agricrithon.*
Legumes	*Atylosia barbata, A. scarabaeoides, A. villosa, Canavalia ensiformis, Mucuna bractearata, vigna umbellate, V. radiata* var. *sublobata, V. pilosa.*
Fibers	*Gossypium aboreum* (primitive types).
Spices and Condiments	*Allium tuberosum, A. sublobatum, Curcuma zedoaria, Alpinia galanga, A. speciosa, Amomum aromaticum, Cucuma amoda, Piper peepuloides.*
Miscellaneous	*Saccharum langisetosum, S. sikkimensis, S. ravennae, Erianchus* spp., *Miscanthus nudipus, M. nepalensis, M. taylorii, Naranga fallax, Camellia* spp.

Source: Arora, R.K. and Nayar, E.R., *Wild Relatives of Crop Plants in India*, National Bureau of Plant Genetic Resources (NBPGR), New Delhi, India, 1984.

It is due to the fragility that the highland ecosystems are extremely vulnerable to extensive changes in land use and developmental activities. When there is disturbance of any degree on the mountains, the adverse changes relating to resource degradation and environmental disruption are very rapid. It is due to the fragility of these ecosystems that the damages done to them are mostly irreversible, or the damage can be healed through ecological regeneration only over a long period. Indiscriminate logging, extensive cultivation, construction activities, road building, mining, megadam construction, and more recently climate change take a heavy toll on the fragile mountains, which is largely attributable to their high degrees of fragility.

ADAPTATION MECHANISMS

Mountain ecosystems pose several constraints for human beings to inhabit there. Extreme climatic variations, acute slopes, extreme fragility, and heterogeneity are quite difficult for the people to adapt to in mountain conditions. However, mountain people have evolved unique adaptation mechanisms to live quite comfortably in the otherwise difficult and inhospitable habitats. Through trial-and-error experiments over several generations, the people—rather mountain people, to be more precise—have modified mountain conditions, for example, through reducing slope steepness through terracing, developing irrigation systems for better agricultural activities and, modifying vegetation composition to fulfill their needs. Apart from altering mountain conditions, the inhabitants also modified their living styles, for example, development of various farming situations and harnessing ecological niches, evolving local means of communication (ropeways, etc.), and exploitation of local resources for house construction, energy sources for warming their houses and cooking foods and, adapting to local food resources and environmental conditions.

The concept of human-adaptation mechanisms would seem to have lost much of its shine if we look at modern development advances: mechanization of land-based livelihoods, infrastructure, overwhelming dependence on markets, network of communication means, availability of technological options, institutional intervention, etc. Nevertheless, an understanding of how traditional highland inhabitants changed the mountain scenario and changed their lifestyles in response to local environmental conditions leads us to ponder over how vegetation composition, ecosystem structure, and introduction of exotic plants took place thanks to large-scale human interventions in the harsher ecosystems. It also helps us understand the rationale of the ongoing climate changes, which are largely owing to gradually accelerated erosion of natural climax forests with associated environmental disruption: catastrophic landslides, excessive soil erosion, biodiversity erosion, twist in hydrological cycles, glacier melt, environmental pollution, erratic weather cycles, and eventually, climate change.

MARGINALITY

Marginality in the context of socioeconomic systems means limitation of the resource base with reference to the "mainstream" situation. Although it is a socioeconomic criterion, in high-altitude human habitats, this prevailing situation has a bearing on plant life. Remoteness, physical isolation, fragility, low productivity of the resource base (Jodha 1993) and, more recently, increasing population, are the major driving forces imposing a sort of geographical, biophysical, and socioeconomic marginality (Singh 2006) on mountain inhabitants.

Such resource-related conditions prevailing in the "marginal" areas compel the local inhabitants to adopt invariably similar patterns of land use involving limited numbers of plant species and genetic varieties, which are of critical socioeconomic importance. The glaring examples reflecting marginality specificity of the mountain habitats are agricultural systems based on typical land use, cropping systems, crop rotations, etc., which vary according to altitude ranges but create a scenario of managed monocultures. The socioeconomic trends in recent years have been greatly influenced by external market demands, which emphasize only a few products and by-products specific to mountain environments (e.g., apples). A compromise with natural diversity is not only visible in private-property resources, such as cultivated lands, but also in respect of common-property resources where socioeconomic activities hardly assign preference to natural heterogeneity. As a result, ecosystem structures and plant species' compositions have changed over time, and this trend is gradually injecting vulnerability into the native ecosystems.

POOR ACCESSIBILITY

High-altitude ecosystems are inaccessibility-ridden in the sense that all areas or parts of the high mountains cannot be accessed easily. High altitude, acute slopes, deep gorges, overall terrain conditions, annual periodic hazards (e.g., snow cover during winter season, landslides and landslips in monsoon season, and storms during summer) impart high degrees of inaccessibility to the mountains. The isolation of human settlements, limited mobility, poor communication, poor linkages with the market system, constraints to developmental activities, limitations to anthropogenic interventions relating to land-use change and natural resource exploitation, etc. are the natural manifestations of the limited accessibility in mountain ecosystems.

Poor accessibility has its bearing on natural resources like forests, rangelands, alpine pastures, and the like, due to restricted anthropogenic activities. This keeps many of the natural resources intact and in perfect natural evolution. Many forests stay virgin and vibrant with their natural functions. Biodiversity in many inaccessible and poorly accessed areas or parts of the mountains is seen flourishing, which is because the same hasn't been exploited like in the readily accessible areas. While the high degree of inaccessibility is regarded as a negative attribute of mountain areas, it is a boon for a variety of plants and the ecosystems they thrive in.

HIGH-ALTITUDE PLANT ECOLOGY

In high-altitude mountains, a very broad range of environmental factors prevails and, therefore, there exists a very wide variety of habitats thanks to continuous interactions among the altitudinal, topological, and geological factors. Factors arising out of altitudinal gradients create specific climatic zones. Topography generates the factors that impart specific terrain structure and conditions. Mountain geology encompasses the factors creating specific edaphic environment (Figure 7.1). Such a diversity of habitats contributes to a unique ecological setting in the mountains of the world. With the specific

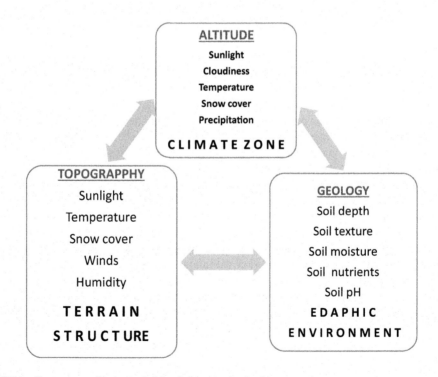

FIGURE 7.1 Factors contributing to habitat heterogeneity in the mountains.

environmental conditions, the diverse habitat's paleorecords reveal that life on Earth began and first nurtured in hospitable environments: warm, sheltered, moist areas. In due course of its evolution, life expanded into relatively less hospitable and eventually into extremely inhospitable environments, such as those experiencing very high or very low temperatures, scarcity of water, and frequent physical disturbances. In the process of occupation of specific environments, the plants had to adapt to those conditions. For example, the plants thriving in the cold-ridden high-altitude environments were those that had morphological and anatomical structures necessary to tolerate low and freezing temperatures. In order to survive freezing temperatures, the plants have to have ultrastructural and molecular specificities that could help them tolerate the freezing conditions and "adjust" their physiology in tune with the habitat specificities. Evolution of the plants' abilities to survive and thrive in extremely cold environments opened the highlands of the world to offer habitats for a variety of plants. Of very high appreciation by botanists are the taxa of the alpine zone in the mountains. Evolutionary strides at high altitudes of the earth have turned mountains into habitats flourishing with extraordinarily rich varieties of plants, with each species encompassing unique characteristics—morphological, anatomical, genetic, and physiological—and adaptation capabilities.

The only "limitation" the high-altitude plants encounter are the environmental "constraints" they survive and reproduce in. Alpine herbs in the high-altitude Himalayas remain embedded under snow for more than six months in a year, and yet they survive well and reproduce normally. Many high-altitude plants can survive even in the "extremes," even if dipped in liquid nitrogen, and yet some can be grown even at an altitude of 6,000 m (Körner 2003), challenging the incredible "extremes." Baffling the survival fitness of a variety of plants amidst the high-altitude extremes, in fact, has been fascinating botanists for decades. Several workers have been working on specialized traits and the physiological uniqueness of these plants. Appreciating the unique traits the high-altitude plants embrace, one can conclude that the incredibly extreme habitats provide specific inputs that these plants "enjoy" for living "normal" lives. The "normal" life of these plants in the "abnormal" habitat is evident from two types of perceived scenarios. One, if the plants from "more favorable" lower altitudes (e.g., flatlands or planes) are shifted to the extreme environment of high altitudes, they would hardly survive. Two, if the plants from the high-altitude extreme environments are shifted to "more favorable" lower altitudes, they will either not survive or will be suppressed by the native plants. The limitation in the context of native plants in high-altitude habitats, in fact, offers favorable "elements," which support only the specialized plants evolved in and well-adapted to these habitats.

There are three mechanisms the plants employ to develop the ability to cope with the specific environmental demands emanating from their habitats (Körner 2003): (1) adaptation or phylogenetic, (2) modification or ontogenic, and (3) modulation or acclimation.

Adaptation mechanisms are evolutionary and non-reversible. Modifications are also non-reversible. Both of these mechanisms are not inherent with the individual plants or their modules (leaves or tillers, for example). Modulation or acclimation mechanisms in response to specific environmental demands, however, are reversible adjustments among individuals in their environments. These are the inherent, but not the exclusive, attributes to help the individuals to survive and sustainably perform in their environments. A plant species or individuals of the species performing well by employing one or all of these adaptive mechanisms would fit into the "survival of the fittest" notion.

Natural selection of the species, however, does not need to employ all the previously mentioned adaptive mechanisms. It rather relies on the genotypic evolution the "naturally selected" species have acquired over millennia. It is through this genotypic attribute that a variety of ecosystems embracing a variety of species all over the planet, ranging from hydrothermal vents in deep ocean floors to alpine zones in the high Himalayas and the Alps, have come into being in the course of natural evolution.

The environment has an overwhelming influence on the morphogenetic characteristics of plant species. Species' populations with particular fitness of life within a particular environment are called ecotypes. The high-altitude plant ecology especially encompasses the concept of ecotypes.

Altitude-specific ecotypes, as Körner (2003) suggests, are only halfway to speciation. Specific plants thriving in high altitudes reflect a higher degree of adaptation compared to those ecotypes that occur at lower or mid-altitudes of the mountains. The photosynthetic efficiency of the high-altitude alpine plants is also comparatively high. High-altitude ecology, thus, sets out several interesting issues worth resolving and establishing rich reserves of ecological knowledge about the incredibly rich repositories of high-altitude ecosystems and ecotypes.

EFFECTS OF ALTITUDES ON THE PHYSIOLOGY OF THE PLANTS

About 8% of the globe's land surface is elevated more than 1,500 m from MSL (Körner 2007). An increase in altitude sets a different set of environmental factors for the plant community, for example, cold climate, low atmospheric pressure, strong winds and, more recently, higher rates of warming (Körner 2003; Rangwala and Miller 2012; Singh et al. 2013; Singh 2018). The environmental conditions prevailing in high-altitude mountains, in essence, are unique on Earth. Generally, high altitudes are less favorable for plants. The biodiversity count, therefore, goes on decreasing with an increase in altitude. It is because compared to lower altitudes, there are fewer plants equipped with the ability to survive in and adapt well to high altitudes. It is due to altitudes that vegetation in the mountains is segmented into a variety of ecological zones. The other critical factors influencing vegetation are geographical latitude and the slope orientation (north slope or south slope/sun slope or shade slope, windy or wind-protected slope). Most of the plant-physiology-related research work in relation with the high-altitude environment is confined to alpine habitats.

Photosynthesis

Photosynthesis, a vital physiological process, is readily affected by changes in environmental conditions. Adaptation to photosynthesis in the alpine plants would depend on three major conditions pervading the high-altitude environment: (1) high light intensity, (2) low temperature, and (3) low CO_2 partial pressure.

High altitudes inhabit the plants that get saturated with greater intensity of light compared to those at lower altitudes and in the plains. As the high altitudes receive greater light intensity, the type of the alpine vegetation is adapted to derive advantage of the "on-the-spot" availability of this energy resource. Earlier experiments conducted by Mooney and Billings (1961) demonstrated that *Oxyria digyna* plants recorded continuous photosynthesis even at 56,000 lux at a 3,740-m altitude, whereas this species' photosynthesis came to saturation at 22,000 lux at a 1,740-m altitude. Tranquillini (1964) has reviewed the experimental findings of Glagoleva, in the Pamir Mountain range in Russia in 1962 and 1963, who observed that light saturation in the high habitats was only at 75,000 to 90,000 lux, arriving at the conclusion that it was 50%–60% maximum field intensities. Glagoleva related high values of her experimental results with the 5–6 layers of thickness of the palisade tissues. The light intensity's effect on photosynthesis, however, is affected by CO_2 concentration. If the CO_2 concentration increases, the light-saturation value also increases.

Alpine plants in the high-altitude environments, for example, in the Himalayan region, remain embedded under a thick layer of snow and stay in dormancy for about six months in a year. Their morphological, cytological, biochemical, and genetic specificities help them flourish even amidst a cryosphere. The growing season of alpine plants begins when the ice melts and night frosts become less frequent during the spring season. It is this season when photosynthetic capacity slowly climbs. With the photosynthetic apparatus involving the ultramicroanatomical structure of the chloroplasts and photoprotection mechanisms associated with alpine plants, they are well adapted to drive photosynthesis at much lower temperatures than their counterparts at lower altitudes and in the warmer climates. Photosynthetic compensation and saturation both require exposure of sun plants to more light. The alpine plants have evolved the ability to make better use of the available light in the high-altitude environments.

As there is enormous variation in the intensity of light and temperature range and immense habitat heterogeneity at high altitude, photosynthetic rates are not uniform throughout a growing season. In the beginning of the growing season when environmental conditions are set for plant growth and new shoots sprout out, there are low net photosynthetic rates due to the rapid growth of new shoots. With a rise in temperatures toward the peak of the summer season, photosynthetic rates increase and attain their peaks during the flowering stage. An increase in the photosynthetic rates, however, is conditioned by the availability of water, a critical input (as well as output) in the process of photosynthesis.

The photosynthetic rate and subsequent growth are determined by the rate of CO_2 uptake by plant leaves. As altitude increases, the CO_2 partial pressure decreases. With other conditions being normal, especially irradiance and water availability, alpine plants are able to assimilate more carbon under low CO_2 partial pressure at high altitudes than those in the habitats near sea level. In his experiments, Tranquillini (1964) found that at 0.4 mg CO_2/L (the approximate CO_2 partial pressure at 3,000-m altitude) the plants collected from a sea-level habitat were capable of only 70% as much photosynthesis as those found in the high-altitude mountains. However, Gale (1972) elucidated that a decrease in the CO_2 availability necessitated for photosynthesis with an increase in altitude is very small and is compensated by a higher CO_2 diffusion rate with increasing altitude. Thus, CO_2 partial pressure at high altitude does not fall to a considerable extent. The prevailing atmospheric temperature and CO_2-uptake characteristics of a plant also determine the CO_2 availability and atmospheric-pressure relationship. The CO_2 assimilation rates at high altitudes, despite the apprehension of low CO_2 partial pressure (Friend and Woodward 1990), have been usually recorded high and even higher than those at low altitudes (Shi et al. 2006; Wang et al. 2017).

Snow cover and cloudy weather impose restrictions on CO_2 uptake. However, it is not precisely the temperature that affects the CO_2 uptake by an overall photosynthetic performance of the plants, but the light regime duration of the season (or leaf longevity). Photosynthesis acclimates to the temperatures prevailing in high-altitude habitats. The photosynthesis–temperature response curve depicts a bell-shaped curve enabling the alpine plants to attain more than 80% of the maximum photosynthesis over the broad temperature range (Figure 7.2).

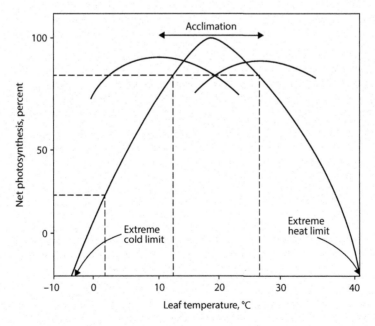

FIGURE 7.2   Photosynthesis–temperature response curve.

At temperature near 0°C, photosynthesis attains 20%–30% of the maximum rate and is halted only when plant leaves freeze at temperatures between −3°C and −6°C.

RESPIRATION

What is the relationship of respiration with temperature? Pisek and Winkler (1958) reported that the high-altitude tree branches near the timberline respire more than their counterparts at lower altitudes during the summer as well as winter seasons, respiration in summer season being 50% more. Thus, the entire temperature range in high-altitude mountains does not affect the plants maintaining respiration rates higher compared to those prevailing at lower altitudes. Since CO_2 evolves through the physiological process of respiration, it could be expected that continuous higher respiration would strike an unfavorable carbon balance. However, it is not the case because the temperatures at high altitudes where those trees grow are nearly always low, as Pisek and Winkler (1958) also suggested, and it is for this reason that the night respiration maintained by the trees in timberline habitats in the mountains was substantially lower than those inhabiting the plain areas (Tranquillini 1964).

The respiration rates of alpine plants are not uniform throughout the growing season. During light hours, CO_2 emission during respiration is masked by CO_2 fixation during photosynthesis. As photosynthesis is enacted only by the green parts of the plants, the other (non-green) parts, viz. roots, rhizomes, flowers etc., continuously emit CO_2 during respiration. And again, the green parts also lose CO_2 during respiration in the dark period ("dark" or "mitochondrial" respiration). With the light being a critical factor, the CO_2 uptake and CO_2 emission in alpine plants, on the whole, is a play in favor of CO_2 uptake. On average, about half of the CO_2 fixed by the plants in photosynthesis goes back to the atmosphere through respiration. Thus, with other favorable environmental factors operating on an alpine habitat, photosynthesis is "dominant" over respiration.

There are different drivers of photosynthesis and mitochondrial respiration. Whereas light is the major driver of the former, temperature is that of the latter. Let us have a look at the respiration–temperature relationship. There are two types of mitochondrial respiration responses to an increase in temperature: spontaneous or short-term response, and long-term response. In the spontaneous response, the respiration rate increases exponentially with increase in temperature, with the average increase being approximately twofold for every 10°K rise in temperature. The Q10 value for the alpine plants has been found to be 2.3. This value in the alpine plants inhabiting the highest altitudes has been recorded between 3 and 5. These "highest" alpine plants perform normal respiration even under extremely lower temperature, which is attributable to the large number of mitochondria in their cells. However, the rates increase rapidly with a rise in temperature with a consequence of rapid CO_2 loss into atmosphere. It is due to the excessive respiratory rates of specialized high-altitude alpine plants in warm temperatures that they cannot survive in warmer conditions at lower altitudes.

The long-term response applies to the conditions when high-altitude plants are moved to lower altitudes, that is, from cold to a warmer temperature zone. Since plants tend to acclimatize to new environmental conditions, their respiratory rates do not increase to a great extent but are maintained at normal rates or only slight increase in respiration.

TRANSPIRATION

Capacity and behavior of transpiration in alpine plants are not fundamentally distinguishable from those at lower altitudes and in the plains. There are differences within the species in the same phytocommunity with regard to transpiration rates. Some alpine plants reduce transpiration when the weather is sunny at midday. Yet there are plants in the same high-altitude habitats that transpire continuously, irrespective of sunshine at noon, showing the transpiration curve with only one peak. Variations in transpiration rates also exist among plant communities, that is, among alpine ecosystems.

Plants on rocky-rubble mountain slopes register the highest transpiration rates; for example, *Rumex* on a clear day has been found losing 12.3 g of water per gram of fresh weight. At noontime, however, in a bid to strike a water balance, these plants demonstrate substantial curtailment in their respiration rates. The water content of the *Rumex* leaves remains almost unchanged even at peak daytime temperatures, and osmotic values are also kept low. Water balance maintained by the plants even with the high-transpiration rate is attributable to the plants' harping on moist earth under the rubble (Tranquillini 1964). *Pinus sylvestris* forests in the alpine conditions regulate transpiration rates with soil water availability and show strong adaptation to soil drought during high-evaporation seasons (Wieser et al. 2014). The root systems of the plants growing on sloppy lands are also quite extensive, helping the plants maintain transpiration rates and water balance.

The snow-bank areas in the alpine zones exhibit moist soil due to continuous snow melt and flowing water into these areas. The major characteristics of the plants growing in such areas serve to maintain low transpiration rates, which are not curtailed during the daytime, even if it is sunny. The osmotic pressure maintained, therefore, is higher than that of the scree plants exhibiting restricted transpiration. A continuous opening of the stomates of the snow-bank plants helps them maintain increased inward flow of CO_2 for biomass production within a short growing (summer) season. Higher inward flow of CO_2 means more evaporative loss. Thus, the plants in these ecosystems have to "bargain" impaired water balance.

Chlorophyll Destruction

Plants in the alpine habitats are generally more photosensitive than their counterparts in lowland habitats. A number of alpine plants are left with extensively destroyed chlorophyll and inhibited chlorophyll synthesis in the presence of light. The upper layers of the palisade layer in the leaves of such plants contain only the faded chloroplasts. Conifers at the timberline, which are especially photosensitive, often suffer from substantial chlorophyll loss during clear sunny days in the winter season. The more intensive the solar radiation the larger the amount of the chlorophyll lost. The needles of the sun-exposed part of the tree canopy lose more chlorophyll than those that are less exposed to solar radiation. The chlorophyll destruction due to intense solar radiation is so severe that conifer needles turn from green to yellow. To protect itself from loss, chloroplasts tend to migrate to the middle of the cell and concentrate around the nucleus, or they withdraw itself into the corners of the fold parenchyma.

The winter season loss of chlorophyll in the sun-exposed conifer needles leads to the reduction in the photosynthetic potential of the high-altitude conifers. In the late-winter season, the branches of the tree that were turned yellow exhibit a negative CO_2 balance. With the onset of the spring season and increase in atmospheric temperature, chlorophyll loss is recouped and the normal photosynthetic potential of the plants is regained.

Resistance to Ultraviolet Radiation

High-altitude habitats are more likely to suffer from ultraviolet radiation. It is due to this fact that alpine plants must also suffer from radiation stress. In an experiment conducted in early 1940s, it was demonstrated that the exposure of alpine plants to artificial ultraviolet radiation led to a slight inhibition of their elongation compared to the plants that received only visible light. However, there was no damage of alpine plants due to ultraviolet light. Exposure to ultraviolet light to some sea-level plants led to their death, whereas the alpine plants were not harmed by the same amount of radiation. The protection of the alpine plants from ultraviolet radiation is attributable to the filtering action of the upper epidermis. Cell sap, not the cell wall, plays a role in the filtration of ultraviolet radiation. Ultraviolet transparency is induced by CO_2 deficiency and darkness, indicating that the ultraviolet radiation is absorbed through photosynthetic products (Tranquillini 1964).

The overall effect of the ultraviolet light on the elongation and growth of plants has not been ascertained. It is also not clear whether this high-energy radiation contributes to promote genetic differentiation of the plants in alpine regions (Körner 2003).

Frost Resistance

Winter drought and freezing become the major causes of damage inflicted on alpine plants especially in late winter. It is often difficult to ascertain which of these two factors is more decisive. Nevertheless, the high-altitude plants are quite resistant to frost damage, which is reflected in the frost hardiness of plant leaves that varies throughout the year, being highest in winter and minimal in summer. It is the temperature history of the plant that determines the degree of the annual frost resistance. An internal annual rhythm created by fluctuating environmental temperature influences the annual frost-resistance curve.

There are, in fact, multiple ways through which a specialized alpine plant combats frost and freezing temperatures. Developing an alternative phenology, morphology, and variable growth forms is helpful for them to elude or lessen the effects of the extremely low temperatures of their environment. The plants can also develop what is known as freezing-point depression by elevating the solute concentration in their cold-exposed tissues. Ice crystallization in plant tissues is also avoided by means of supercooling. Plants also apply the strategy of dehydrating their cells by causing the movement of water in their cells to their intercellular spaces. In this way, ice formation in the cell is shifted to other locations, and frost damage is avoided.

When all these tactics of alpine plants to prevent frost damage do not work, they resort to use the ultimate "weapon": exercising their ability to repair and replace their damaged organs by placing their meristems below ground (Körner 2003; Hacker and Neuner 2008), thus escaping above-ground freezing temperatures.

Water Balance and Avoidance of Desiccation

In the winter season, before the alpine plants are covered with snow, they have to encounter soil freezing. Under that situation, if the plants had faced a water deficit in the summer, their leaves would have absorbed water vapor from the surrounding air, and the plants would cope with the water deficit. Subsequently, when the alpine plants wear a thick cover of snow and the atmosphere around them is water saturated, there is no water loss to be encountered by the plants. The alpine region turns into a cryosphere, thus, helping the native plants maintain a water balance and protect them from desiccation. With the onset of the spring season, the snow begins melting gradually, and plants begin absorbing water from the moist soils, enabling them to be saturated with their water content and their osmotic values drop.

The large evergreen trees in the timberline zone "enjoy" their well-equalized water balance in the winter season. The water balance in the trees is maintained and desiccation is avoided due to three reasons:

1. Stomata in the leaves of these trees remain closed for the entire season, ensuring a minimum loss through transpiration.
2. Deep-root system of the large trees derives an advantage from the water from a deeper, unfrozen soil profile.
3. The large trees have enough water reserves in their trunks and branches.

In summer season too, there is hardly a danger for the alpine vegetation to face drought conditions. The water loss of the plants is generally lower than the annual precipitation. There are chances that there is always enough precipitation with equal water distribution in alpine habitats. Further, snow melt in the summer helps to maintain moisture in the soil environment. The osmotic value of the alpine plants during the summer season is also held somewhat higher compared to their lowland counterparts.

Alpine habitats are generally characterized by variable-water availability. These habitats are more favorable for the plants carrying the properties of desiccation tolerance. As the alpine habitats have an abundance of bryophytes and lichens (Austrheim et al. 2005), plants of these groups exhibit a high degree of desiccation tolerance. One of the "strategic" measures that the alpine succulent plants adopt is CAM (crassulacean acid metabolism) photosynthesis, which is a water-economic pathway of carbon fixation.

GROWTH AND DEVELOPMENT

The most common plants in alpine ecosystems include perennial herbs with well-developed rhizomes and/or root systems. The below-ground system maintains carbohydrate reserves during the winter season when growth of the plants almost ceases. The root/rhizome reserves of the carbohydrates serve as energy sources for the plants to develop new shoots during the spring/summer season when vegetative dormancy breaks down. In Saxifrages (the alpine plants belonging to family Saxifragaceae), energy reserves in terms of carbohydrates and lipids are concentrated in their succulent leaves rather than in their roots. As the summer season comes to a close, accompanying the spell of short photoperiod, the alpine plants resume vegetative dormancy, forming perennating buds containing nutrients enough to sustain life during the winter season.

It takes a few years for the perennial alpine herbs to be established in their ecosystems. The seedling process is pretty slow, and vegetative reproduction is the dominant mode of plant life in alpine habitats. During the first year of a plant, most of the energy it accumulates through photosynthesis is used for the growth and establishment of the root system, which is a necessity for the plants to avoid desiccation. A fraction of the energy is used up for the metabolism during the period of vegetative dormancy. In the first year, the plant usually produces cotyledons, but some true leaves are also produced. Thus, the successive perennial journey of an alpine plant in its habitat goes on up to the realization of its genetic potential amidst interactions with abiotic and biotic factors operating in a high-altitude environment.

Extreme low temperatures pervading alpine regions reduce photosynthetic rates directly as well as indirectly. This occurs in the low-temperature response to photosynthetic enzymes involved in the Calvin cycle, such as RuBisCO. RuBP carboxylation influenced by low temperatures also has a role to play in limiting photosynthetic rates in cold-tolerant species (Yamori et al. 2010). Reduced stomatal conductance is another cause of a reduced photosynthetic rate (Allen et al. 2002). However, the alpine plants have their own CO_2 assimilation potential, which has an edge over that of non-alpine plants.

The long winter season turning the alpine region into a cryosphere, late sprouting and early leaf fall in "summer green" plants, and a lengthy dormant period of "evergreen" plants reduce the active growing period (of about three months) for the alpine vegetation. This, therefore, can be expected that the alpine plants, unlike the plants in non-alpine ecosystems, have to synthesize and assimilate biomass within a limited period. In other words, their biomass production and assimilation economy are determined by a short growing period.

The results that poured in from earlier investigations between 1940s and early 1960s, as reviewed by Tranquillini (1964), revealed that the high-altitude plants performed more actively and assimilated more strongly than those at lower altitudes. It was found that under CO_2 saturation, the maximum potential net photosynthesis of various plant species was up to two times higher at an altitude of 3,860 m than at 2,350 m in the mountains of Pamir. The altitude and potential net photosynthesis relationship, however, is not linear because no further increase in the net photosynthesis was recorded at 4,780 m in the Pamir. The maximum value of assimilation found was 200 mg/g dry weight/hr in *Primula algida*, while other species (e.g., *Stilphonophleum anthoxanthoides*) recorded much less: 30 mg/g dry weight/hr. Considering that nearly half of the plant biomass on dry-matter basis is comprised of carbon, it was found that 80 mg of carbon were added every day to each gram of carbon already present.

The freezing night temperature tends to influence CO_2 assimilation rate. Huang et al. (2016) conducted experiments on alpine evergreen broadleaf tree species *Quercus guyavifolia* at 2,500–4,300-m altitude in China during the winter season and found that the inhibited rate of photosynthesis due to freezing night temperatures, photorespiration, and alternate electron flow play important roles in the regulation of photosynthetic electron flow, which, in turn, helps protect the photosynthetic apparatus from photodamage.

Alpine plants, despite a short growth period, are left with adequate amounts of assimilate and express a positive growth status of their existence. The high-altitude mountain habitats, in fact, must be appreciated for their specific plant communities excelling in their ability to utilize light and realize their growth potential.

HIGH ALTITUDES: COOLING BREEZE FOR COOLING THE MAINSTREAM WORLD

Mountain ecosystems are extended over about 24% of the earth's land surface and are home to about 24% of the world's population (12% in mountain habitats and 12% in the immediate proximity). In addition, mountain ecosystems support an overwhelming proportion of humanity everywhere by providing valuable goods and vital services. They serve as water towers for humanity. All big rivers on the planet exist owing to their origin in the mountains. Plain areas receive river waters for drinking, irrigation, and industrial purposes thanks to the mountains. Can you imagine rivers and river systems without mountains? And, of course, the mountain ecosystems are rich natural repositories of biological resources. The fertility of the plain areas is largely attributable to the mountains.

Mountain ecosystems build up their own microclimate and phenomenally influence the climate across the planet at many scales. Despite being in tropical regions of the globe, the highland landscapes, such as the Himalayas, are cooler and wetter than the plain areas in the same geographical region. It is, as we have already discussed, due to altitude–temperature relationships. In the troposphere, where mountains of the planet are situated, there is about a 6.5°K drop in temperature with every 1-km increase in altitude up to 11,000-m altitude, as per the environmental lapse rate (ELR) we have discussed elsewhere. There is an alternative explanation of the cooling airs on mountaintops: air goes on becoming thinner and less able to absorb and retain heat along increasing altitude. The lower the temperature the less the evaporation, which means more moisture in the air. The light- and low-pressure air at high altitudes expands and goes on becoming cooler and spells a cooling effect over the mountain landscapes.

The essence is that the climate regime in the mountain ecosystems is cooler than that of the surrounding flatlands. The warmer climate in the plains of the tropical regions, such as in Asia, does not exert much impact on the high-altitude mountain climate. The reverse, however, is true; that is, the mountains exert a certain degree of cooling effect on the climate of nearby flatland areas. Mountains are often alleged to be "harsh" habitats. But from the viewpoint of climate, the highland habitats are very soothing, especially in the hot tropical regions.

Mountains, because of low temperatures, generally receive more rains than the plains. Winds blow over the land with moist air. When mountains obstruct the winds, it rises up and gets cooler due to lower temperatures at high altitude. Since the cool air carries less moisture than the warm air, the consequence is precipitation. Mountain weather can change dramatically even during short periods. Very high peaks of the mountains, even near the equator, are covered with snow throughout the year. The Himalayas in Asia, outside the polar regions of the earth, are the largest home to snow and ice. Therefore, the Himalayas are sometimes referred to as the "Third Pole." Glaciers in the mountains serve as the perennial sources of rivers. Himalayas are the sources of some of the largest river systems of the earth. Himalayan glaciers cover about 3 million ha, or 17% of the global mountain area—the largest bodies of ice outside the polar caps. The total area of Himalayan glaciers is 35,110 km^2. The total ice reserve of these glaciers is 3,735 km^3, which is equivalent to 3,250 km^3 of fresh water. Himalayan mountains are sources of the nine giant river systems of Asia: the Indus, the Ganga, the Brahmaputra, the Irrawaddy, the Salween, the Mekong, the Yangtze, the Yellow, and the Tarim. The Himalayan Mountain region serves as the water lifeline for more than 500 million inhabitants of the region, or about 10% of the total regional human population (IPCC 2007).

Descending from the highlands and choppily flowing through vast plain areas, these rivers and river systems go on giving a cooling touch to the surrounding environment.

In our present times, mountain ecology is not as healthy as it is expected to be. The recession of glaciers, catastrophic mining, rampant deforestation, indiscriminate road construction, disastrous blasting, devastating bulldozing, increasing urbanization, intensive tourist activities, extensive terracing of mountain slopes, etc. are the extremely destructive changes being increasingly witnessed in many mountain areas of the world, especially in the Himalayas, which have turned into the battleground under prevailing tensions among neighboring countries. The extreme northwest part of the Hindu Kush-Himalayas has turned into a hot spot of terrorism. Such pressures on the mountain ecosystems are disrupting mountain ecology and curtailing vital attributes emanating from specific ecosystem functions.

Mountains are the islands in the sky. Their ecological well-being is vital for the ecological well-being for the mainstream world of the plains. These ecosystems are the natural global heritage. Therefore, the entire world should come to the fore to resolve mountain crises and mountain dilemmas. The political consensus among the governments sharing the mountains and highlands needs to be urgently realized. Border areas in the mountains should be declared biological corridors. These biological corridors will enhance ecosystem services for the common benefit of all the mountain-sharing countries. An international law declaring mountains of the world as "Peace Zones" will prove phenomenal for ameliorating the ecology of the highland ecosystems, a state that would give a healing impact to the changing global climate.

Under the suggested peaceful conditions and adoption of appropriate policies and programs in tune with the mountain perspective, what will be the key to the vibrant ecology of the world's mountains? It is the natural forests. Mountains must be laden with natural forests. For this, protection and conservation of existing natural forests, ecological regeneration of the existing degraded forests, afforestation involving native species on a large scale, widespread adoption of agroforestry systems in cultivated/terraced lands, conservation of natural pastures/rangelands, etc. will be the part of the strategies to evolve healthy and vibrant ecosystems in the highlands. The largest possible geographical areas of the mountains wearing dense vegetation cover will be phenomenal for soil, water, and biodiversity conservation, for regulating the water cycle, and for reviving the climate conducive to a healthy and life-enhancing environment.

Despite their vital and phenomenal contributions, the mountains of the world are just regarded as the marginal areas and are seldom counted as sources of global development. In our minds, the plains areas of the lands constitute the mainstream of the world. However, the "marginal world" holds the key to the well-being of the "mainstream world."

The climate mountains build up in the tropical region, like in the Himalayas, ranges from the arctic type in the cryosphere created by glaciers in the upper reaches to the alpine type between the timberline and glaciers, to the temperate in the middle range, to the subtropical in the lower valleys and foothills, to the tropical a few kilometers away from the foothills. In this way, the cool breezes blowing over the mountain ecosystems are one of the best bets for cooling the earth and, thus, in building up a climate-smart planet.

SUMMARY

High-altitude ecosystems of the world are markedly distinguishable from those in the plains or flatlands and are characterized by six specificities, viz. diverse ecological niches, high degree of biodiversity, fragility, adaptation mechanisms, marginality, and limited accessibility. The atmosphere prevails up to an altitude of 500 km and its composition and temperature change along altitudes. Accordingly, the atmosphere is categorized into five segments or zones: troposphere, stratosphere, mesosphere, thermosphere, and exosphere. In the troposphere and mesosphere, the temperature goes on decreasing with an increase in altitude, whereas in other atmospheric layers, the temperature increases along with an increase in altitude. A graphical representation of altitude vs. temperature assumes a Σ shape.

In high-altitude mountains, a very broad range of environmental factors prevails, and therefore, there exists a very wide variety of habitats thanks to continuous interactions among the altitudinal, topological, and geological factors. Many high-altitude specialized plants can survive even in the extremely low temperatures and yet some even can grow at an altitude of 6,000 m.

The three mechanisms that plants employ to develop the ability to cope with the specific environmental demands emanating from their habitats are: (1) adaptation or phylogenetic, (2) modification or ontogenic, and (3) modulation or acclimation. Adaptation mechanisms are evolutionary and non-reversible. Modifications are also non-reversible. Modulation or acclimation mechanisms in response to specific environmental demands, however, are reversible adjustments among individuals in their environments.

The environmental conditions prevailing in high-altitude mountains, in essence, are unique on Earth. Generally, high altitudes are less favorable for plants. The other critical factors influencing vegetation are geographical latitude and the slope orientation (north slope/shade slope or south slope/ sun slope, or windy or wind-protected slope). Adaptation to photosynthesis in the alpine plants would depend on three major conditions pervading the high-altitude environment: (1) high light intensity, (2) low temperature, and (3) low CO_2 partial pressure. High altitudes inhabit the plants that get saturated with a greater intensity of light compared to those at lower altitudes and in the plains. As the high altitudes receive greater light intensity, the alpine vegetation is adapted to derive the advantage of the "on-the-spot" availability of this energy resource. The alpine plants have evolved the ability to make better use of the available light in the high-altitude environments. The entire temperature range in high-altitude mountains does not affect the plants maintaining respiration rates higher compared to those prevailing at lower altitudes. Respiration rates of alpine plants are not uniform throughout the growing season. In the light hours, CO_2 emission during respiration is masked by CO_2 fixation during photosynthesis. There are different drivers of photosynthesis and mitochondrial respiration. Whereas light is the major driver of the former, temperature is that of the latter.

Capacity and behavior of transpiration in alpine plants is not fundamentally distinguishable from those at lower altitudes and in the plains. The major characteristics of the plants growing in moist high-altitude areas serve to maintain low transpiration rates, which are not curtailed during the daytime even if it is sunny.

Plants in the alpine habitats are generally more photosensitive than their counterparts in lowland habitats. High-altitude habitats are more likely to suffer from ultraviolet radiation and so do the alpine plants. Developing alternative phenology, morphology, and variable growth forms are helpful for the alpine plants to elude or lessen the effects of extremely low temperatures of their environment. The plants can also develop freezing-point depression by elevating the solute concentration in their cold-exposed tissues. The water balance in the trees is maintained and desiccation avoided due to three reasons: (1) stomata in the leaves of these trees remain closed for the entire season ensuring minimum loss through transpiration; (2) deep root system of the large trees derives an advantage from the water from a deeper unfrozen soil profile; and (3) the large trees have enough water reserves in their trunks and branches.

Mountain ecosystems build up their own microclimates and phenomenally influence the climate across the planet at many scales. Despite being in tropical regions of the globe, the highland landscapes, such as the Himalayas, are cooler and wetter than the plain areas in the same geographical region, which is attributable to altitude–temperature relationships. In the troposphere, where the mountains of the planet are situated, there is about a 6.5°K drop in temperature with every 1-km increase in altitude up to 11,000 m, as per the ELR. The climate regime in the mountain ecosystems is considerably cooler than that in the surrounding flatlands. The warmer climate in the plains of the tropical regions, such as in Asia, does not exert much impact on the high-altitude mountain climate. The reverse, however, is true; that is, the mountains exert a certain degree of cooling effect on the climate of nearby flatland areas.

The climate mountains build up in the tropical region, like in the Himalayas, ranges from the arctic type in the cryosphere created by glaciers in the upper reaches to the alpine type between the timberline and glaciers, to the temperate in the middle range, to the subtropical in the lower valleys and foothills, to the tropical a few kilometers away from the foothills. In this way, the cool breezes blowing over the mountain ecosystems are one of the best bets for cooling the earth and, thus, in the building up a climate-smart planet. An ecologically sound management of the high-altitude environments will impart a cooling effect on the general climate of the mainstream world.

REFERENCES

Allen, D. J., Ratner, K., Giller, Y. E., Gussakovsky, E. E., Shahak, Y. and Ort, D. R. 2002. An overnight chill induces a delayed inhibition of photosynthesis at midday in mango (*Mangifera indica* L.) *J. Exp. Bot.* 51: 1893–1902.

Arora, R. K. and Nayar, E. R. 1984. *Wild Relatives of Crop Plants in India*. New Delhi, India: National Bureau of Plant Genetic Resources (NBPGR).

Austrheim, G., Hassel, K. and Mysterud, A. 2005. The role of life history traits for Bryophyte community patterns in two contrasting alpine regions. *Bryologist* 108 (2): 259–271.

Friend, A. D. and Woodward, F. I. 1990. Evolutionary and physiological responses of mountain plants to growing season environment. In: Bagon, M., Fitter, A. H. and Mcfadyen, A. (eds.) *Advances in Ecological Research*, Vol. 20. London, UK: Academic Press, pp. 59–124.

Gale, J. 1972. Availability of carbon dioxide for photosynthesis at high altitudes: Theoretical considerations. *Ecologist* 53 (3): 494–497.

Hacker, J. and Neuner, G. 2008. Ice propagation in Dehardened alpine plant species studied by infrared differential thermal analysis (IDTA). *Arct. Antarct. Alp. Res.* 40 (4): 660–670.

Huang, W., Hu, H. and Zhang, S.-B. 2016. Photosynthesis and photosynthetic electron flow in alpine evergreen species *Quercus guyavifolia* in winter. *Front. Plant Sci.* 7: 1511. doi:10.3389/fpls.2006.01511.

IPCC. 2007. *Climate Change 2007: Impacts, adaptation and vulnerability—Contribution of Working Group II to the Fourth Assessment Report of the IPCC*. Geneva, Switzerland: IPCC.

Jodha, N. S. 1990. *Mountain Agriculture: Search for Sustainability*. Mountain Farming Systems Division Discussion Paper No. 2. Kathmandu, Nepal: ICIMOD.

Jodha, N. S. 1993. A framework for integrated mountain development. In: Singh, V. (ed.) *Eco-Crisis in the Himalaya: Causes, Consequences and Way Out*. Dehradun, India: IBD, pp. 1–35.

Kömer, C. 2003. *Alpine Plant Life: Functional Plant Ecology of High Mountain Ecosystems*. Berlin, Germany: Springer. 351 p.

Körner, C. 2003. *Alpine Plant Life: Functional Plant Ecology of Plants*. New York: Springer, 349 p.

Körner, C. 2007. The use of "altitude" in ecological research. *Trends Ecol. Evol.* 22: 569–574.

Mooney, H. A. and Billings, W. D. 1961. Ecology of Arctic and alpine populations of *Oxyria digyna*. *Ecol. Monographs*. 31: 1–29.

Partap, T. 1993. Biological diversity issue and sustainable mountain development. In: Singh, V. (ed.) *Eco-Crisis in the Himalaya: Causes, Consequences and Way Out*. Dehradun, India: IBD, pp. 78–122.

Pisek, A. and Winkler, E. 1958. Assimilation and respiration der fichte in verschiedener Höhenlage und der Zirbe an der Waldgrenze. *Planta* 51: 518–543.

Rangwala, I. and Miller, J. R. 2012. Climate change in mountains: A review of elevation-dependent warming and its possible causes. *Clim. Change* 114: 527–547.

Shi, Z., Liu, S., Liu, X. and Centritto, M. 2006. Altitudinal variation in photosynthetic capacity, diffusional conductance and $\delta^{13}C$ of butterfly bush (*Buddleja davidii*) plants growing at high elevations. *Physiol. Plant.* 128: 722–731.

Singh, V. 1998. *Draught Animal Power in Mountain Agriculture: A Study of Perspectives and Issues in Central Himalayas, India*. Mountain Farming Systems Discussion Paper 98/2. Kathmandu: Nepal: International Centre for Integrated Mountain Development (ICIMOD).

Singh, V. 2006. Marginality of mountain farmers and strategies to cope with. In: Rawat, M. S. S. (ed.) *Resource Appraisal, Technology Application and Environmental Challenges in Central Himalaya*. Srinagar, Garhwal: Department of Geography, HNB Garhwal University.

Singh, V. 2018a. *An Analysis of Mountain Agro-Ecosystems*. Beau Bassin, Mauritius: LAP Lambert. 64 p.

Singh, V. 2018b. Food Security amidst the climate change scenario: Perspectives, issues and opportunities in mountain agriculture. *Nutr. Food Sci. Int. J.* 6 (2): 001–009.

Singh, V., Nautiyal, N. and Rastogi, A. 2013. Callous development and climate disruption: Environmental management for healing the earth and restoring climate pattern. In: Nautiyal, N., Rastogi, A., Kumar, B. and Singh, V. (eds.) *Environmental Management: Meeting the Challenges of Climate Change*. New Delhi, India: SSDN, pp. 11–20.

Tranquillini, W. 1964. The physiology of plants at high altitudes. *Annu. Rev. Plant Physiol.* 15: 345–362.

Wang, H., Prentice, I. C., Davis, T. W., Keenan, T. F. and Wright, I. J. 2017. Phytosynthetic responses to altitude: An explanation based on optimality principles. *New Phytol.* 213: 976–982.

Wieser, G., Leo, M. and Oberhuber, W. 2014. Transpiration and canopy conductance in an inner alpine Scots pine (*Pinus sylvestris* L.) forest. *Flora* 209 (9): 491–498.

Yamori, W., Noguchi, K., Hikosaka, K. and Terashima, I. 2010. Phenotypic plasticity in photosynthetic temperature acclimation among crop species with different cold tolerances. *Plant Physiol.* 152: 388–399.

WEBSITES

http://www.geo.uzh.ch/microsite/alpecole/static/course/lessons/02/02.htm
http://www.geo.uzh.ch/microsite/alpecole/static/course/lessons/08/08.htm
http://www.primaryhomeworkhelp.co.uk/mountains/climate.htm
http://www.springer.com/978-3-540-00347-2
https://en.wikipedia.org/wiki/Thermosphere
https://en.wikipedia.org/wiki/Troposphere
https://imagine.gsfc.nasa.gov/science/toolbox/emspectrum1.html
https://scied.ucar.edu/shortcontent/thermosphere-overview
https://spaceplace.nasa.gov/stratosphere/en/
https://spaceplace.nasa.gov/thermosphere/en/
https://www.iucn.org/commissions/commission-ecosystem-management/our-work/cems-specialist-groups/mountain-ecosystems
https://www.ncbi.nlm.nih.gov/pmc/articles/PMC4959566/
https://www.siyavula.com/read/science/grade-10/electromagnetic-radiation/11-electromagnetic-radiation-03
https://www.un.org/press/en/2002/sgsm8464.doc.htm

8 Stress Physiology

Stress-tolerant species predominate under conditions of low disturbance and high stress.

J. P. Grime

STRESS TO LIFE AS AN INEVITABLE PHENOMENON

Stress implies some adverse effects on the physiology of organisms induced by environmental factors, abiotic as well as biotic. Grime (1977) defined stress in relation to plants as "external constraints that limit the rate of dry-matter production, or growth, of all or part of the vegetation." The "external constraints" arise out of negative or adverse environmental factors, including pollution, that induce physiological responses in living species or alter the structure and functioning of an ecosystem.

Life on Earth often encounters stress conditions. Stress is a complex phenomenon and yet almost inevitable for organisms and their communities. In some cases stress is occasional, in some frequent, and in yet other cases it is a permanent feature of ecosystems. Environmental conditions are not static. Operating environmental factors are also varying. Our biosphere is dynamic and tends to be in the state of equilibrium. Stress conditions emanate from unfavorable environmental conditions. Unfavorable conditions might be spatial or temporal in nature. In some ecosystems all the time and in some ecosystems at a certain point of time, some or all organisms experience and tend to combat stress conditions and attempt to adapt to those unfavorable conditions.

In order to carry on life processes, such as maintenance (of body structures, body temperature, functions, etc.), transportation of nutrients to different tissues, cellular biosynthesis, capacity to grow and replicate, etc., a continuous input of energy to the organisms is a precondition. Required energy for the organisms is obtained by means of nutrition. The food energy remains in flow throughout the bodies of all organisms throughout their life span. However, the abiotic/meteorological (excessive radiation, temperature, humidity, drought conditions, etc.) and biotic (diseases, pollen grains, mites, insects, pathogens, etc.) environmental factors also play a decisive role in the maintenance, functioning, and sustenance of the organisms, as these factors influence the thermodynamics of the environment and, therefore, the physiology of the organisms. The energy systems of the organisms—both internal and external—strike a steady state that results in a meta-stable condition called homeostasis.

An alteration in the amount of a natural factor even of a smaller measure may affect and even disrupt the homeostasis. Biological stress is, thus, caused by the environmental modulation of homeostasis. The environmental modulation of stress, or biological stress, is an inevitable phenomenon, and all organisms in the biosphere have to bear stress of some kind every now and then, which might be spatial or temporal in nature.

Life on Earth, in essence, faces stresses in several ways. However, at the same time, organisms tend to attain homeostasis and, consequently, to be in the state of acclimatization and adaptation. We can also infer that stresses, homeostasis, acclimatization, and adaptation go on the same axis of the phenomenon of life. This phenomenon is influenced and controlled by the physiological changes within organisms.

SOLAR-RADIATION STRESS

Photoautotrophs, that is, the green plants, depend on solar radiation for their survival and sustenance. It is the visible light or photosynthetically active radiation that drives photosynthesis. However, a plant is exposed to radiation of various wavelengths. The long-wave-energy radiation (infrared) carries a less amount of energy and cannot drive photosynthesis. On the other hand, the short-wave-energy wavelength of the electromagnetic spectrum, especially the ultraviolet (UV) radiation, carries an excessive amount of energy that can inhibit cellular processes by damaging membranes, proteins, and nucleic acids.

In the visible range of the solar spectrum, irradiance far above the light-saturation point of photosynthesis often causes light stress leading to photoinhibition (Figure 8.1) by disrupting the chloroplast structure and reducing photosynthetic rates.

The excess of light causes direct damage of the D1 protein that occurs as a result of excess light excitation at the PSII reaction center and consequent inactivation of the same. Exposed to excessive solar radiation, pigments in plant leaves absorb excessive light energy thereby producing excessive electrons far more than the availability of $NADP^+$ to act as an electron sink at PSI. Excess electrons produced by PSI lead to the production of the reactive oxygen species (ROS), especially the superoxide (O_2^{e-}). Oxidative damage of proteins, lipids, DNA, and RNA is done by the ROS, including superoxide, the low-molecular-weight compounds. This particularly happens when they are produced in excess. ROS in excess amounts puts the plant exposed to overdoses of radiation under oxidative stress, leading to the disruption of cellular and metabolic functions and eventually to cell death.

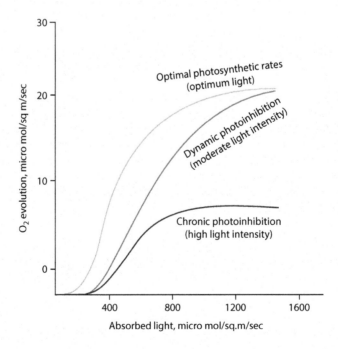

FIGURE 8.1   Changes in the light-response curves of photosynthesis caused by photoinhibition.

TEMPERATURE STRESS

Optimal growth and overall development of an organism is set within a range of temperatures. Some species, for instance, have a narrow range while some have a broad range of temperatures in which they flourish with their natural traits. Beyond a defined temperature range, the species would experience damages of various sorts depending on the magnitude and duration of temperature fluctuation. In other words, an organism will be under stress if the temperature fluctuates beyond a range in which it is unable to maintain its normal physiological state. The stress is mainly due to: (1) high temperature, (2) low temperatures above the melting point, and (3) low temperatures below the freezing point.

In many cases, some plants survive well, defying the temperature range. For example, dry seeds of some plants can tolerate temperature as high as 120°C. Pollen grains of some plant species survive at 70°C.

Most of the C_3 and C_4 plants in suitable habitats, with plentiful water accessibility, are capable of maintaining their leaf temperature well below 45°C, which is due to evaporative cooling despite high ambient temperatures. High leaf temperature coupled with minimal evaporative cooling causes some degree of heat stress. With midday, the bright sunlight and consequent stomata closure to cope with soil-water deficit and high relative humidity reducing the gradient driving evaporative cooling, the leaf temperature can rise to 4°C–5°C above ambient temperature. Leaf temperatures increase above the ambient temperatures attributable to bright sun shining are clearly revealed among drought-ridden plants.

EFFECTS ON PHOTOSYNTHESIS AND RESPIRATION

Both photosynthesis and respiration are readily affected by temperature stress. Photosynthesis is more sensitive to temperature fluctuations than respiration. High temperatures inhibit rates of photosynthesis more than those of respiration. Enzymes in the chloroplasts involved in photosynthesis, viz., RuBisCO, RuBisCO activase, phosphoenolpyruvate (PEP) carboxylase, and NADP-G3P dehydrogenase go unstable at high temperatures. However, the temperatures at which these enzymes become denatured and go inactive and unable to play any role in photosynthesis are markedly higher than at which rates of photosynthesis start declining. This reveals that the early stages in heat injury leading to reduce photosynthetic rates are more directly related to the heat-induced changes in membrane properties and to uncoupling of the energy transfer mechanisms taking place in the chloroplast.

High temperatures strike an imbalance between photosynthesis and respiration, which is attributable to the deleterious effects of the heat on these two vital processes. The compensation point of an individual plant varies for the "shade leaves" and "sun leaves" being lower for the former and higher for the latter. Temperature stress thus would, naturally, lead to reduced production of photosynthate, which is also attributable to a reduction in the leaf canopy area, stomatal closure, and regulation of assimilate partitioning.

EFFECT ON MEMBRANES AND ENZYMES

An environmental factor potent enough to alter membrane properties is also capable of disrupting cellular processes. Plant membranes are made up of a lipid bilayer interspersed with proteins and sterols. H^+-pumping ATPases, carriers, and channel-forming proteins regulating ion transport and solutes are readily affected by physical properties of lipids. In other words, specific actions of the integral membrane proteins are quite sensitive to alteration in the normal properties of the lipid constituent of the cell membrane. Exposure of the plants to high temperatures results in an increase in membrane lipid fluidity coupled with a decrease in hydrogen-bonding strength. There are electrostatic interactions between polar groups of proteins in the aqueous phase, which are also disrupted by high temperatures.

High temperatures knock down three-dimensional structures of proteins, including enzymes, rendering them inactive. Membrane structure and composition get modified when plant tissues are exposed to high temperatures. This state leads to the leakage of ions from the cells. When in the state of denaturation thanks to high temperatures, proteins cause serious problems within cells.

EFFECTS OF FREEZING TEMPERATURES

Freezing temperatures in temperate regions of the world push plants into stress for an extended period of an annual cycle. Low temperatures cause damage to the plants in several ways, and extremely low temperatures for an extended period might even be lethal to the plants. At below-freezing point, water assumes crystal forms within as well as outside plant cells. Shearing of cell organelles and cell membranes takes place after intercellular crystallization of the water. External crystal formation might not be as much harmful as internal. However, external crystallization causes cellular dehydration, which is no less harmful for the plants. With freezing water, water potential in the apoplast decreases and that in the symplast increases. This reverse water-potential gradient allows water to move from the symplast to apoplast causing cellular dehydration.

Extracellular water crystallization does not significantly affect the already dehydrated seeds and pollen grains. In hardy plants, internal freezing of water does not prove lethal. It is because the freezing first takes place in intercellular spaces and xylem vessels from where ice can propagate, and, with a slight increase in temperature, tissues can recover. With exposure of plants to freezing temperatures for longer periods, there is physical destruction of cell membranes and severe dehydration becomes imminent.

STRESS OF OXYGEN DEFICIENCY

Plant roots in terrestrial ecosystems get adequate supplies of oxygen necessary for aerobic respiration. In well-drained and well-structured soils, there are gas-filled pores through which gases, including oxygen, diffuse several meters deep and become available to plant roots. Oxygen in deeper soil layers can be compared with that in the humid air. However, a continuous supply of oxygen in adequate amounts to plant roots is frequently impaired in the flooded and waterlogged soils. If the soils are not well-drained and if irrigation is excessive, then water fills up soil pores and movement of oxygen in the gaseous phase is slowed down considerably.

When the plants are dormant and ambient temperatures are low, oxygen depletion in standing water is slow, and such a state of the environment is not very harmful for the terrestrial plants. However, when the plants are in an active phase and temperatures are higher ($>20°C$), oxygen consumption by soil microflora and fauna and by the plants themselves increases to the extent that the soil in stagnant water gets completely deprived of oxygen. Under such conditions, the plants come into stress imposed by oxygen deficiency.

Exposure of the plants to oxygen-deficiency (hypoxia) conditions or to complete oxygen absence (anoxia) is not uncommon on our planet. And so is the phenomenon of anaerobic stress experienced by the plants. Such conditions often lead to widespread damage and even to death of wild flora and cultivated crops. Such a situation often comes out as a jolt to the economy. Vast areas in many countries are occupied by wet and flooded soils. Plants occurring in such areas frequently suffer from hypoxia and anoxia. Plants suffering from such states of their environment (wet and flooded soils) most often experience damage to their roots and seeds (Maltby 1991), causing loss to the economy as well as to the ecology. If the O_2 solubility and diffusion rates in water are low, the availability of the same to the plants is reduced (Armstrong 1979). Oxygen deficiency to a critical state puts the plants and the plant communities under severe stress.

Compactness of the soil may also create anaerobic conditions. Soils in the vast areas in temperate regions of the earth face acute O_2 scarcity in autumn and the winter season when they are covered by an ice crust impermeable to air. In the era of climate change, larger areas in the polar region and in the foothills of the Himalayas will be flooded by melting ice. Soils, and subsequently plants, in such areas will also face stress of O_2 deficiency (Sachs and Vartapetian 2007).

And yet, there are numerous anthropogenic causes of creating O_2 deficiency. These include excessive irrigation of cultivated fields that renders the soils deficient of oxygen. Long-term storage of fruits, vegetables, and food grains also causes O_2 deficiency.

Accumulation of toxic chemicals due to degradation of organic and inorganic compounds also takes place under anaerobic conditions in the excessively wet and flooded soils. So plants thriving on such soils might also suffer from such toxic chemicals.

As molecular oxygen is a source to generate energy in the aerobic organisms like plants, its deficiency causes damage to a great extent due to which the plants perish, or else, they adapt to extreme conditions of O_2 deficiency. The only glaring exception is rice, one of the dominant rainy-season crops in South Asia that flourishes well in excessively irrigated/flooded soils. Though rice tolerates the conditions of anaerobic soils, this crop has also been found suffering from O_2 deficiency, especially when its seedlings are completely submerged (Jackson and Ram 2003) as is the common scenario in East and South Asia with plentiful water sources for irrigation.

MOLECULAR GENETICS ASPECTS OF OXYGEN-DEFICIENCY STRESS

An increase in the alcohol dehydrogenase (ADH) activity in the maize seedlings submerged in water was observed long back by Hageman and Flesher (1960). Later on, Schwartz (1969) revealed that ADH activity was essential for the survival of maize seeds and seedlings during submergence in water. He also showed that the seedlings survived for about three days of anoxia, but the ADH nulls survived only a few hours of flooding. Being a major terminal enzyme of fermentation in plants, ADH is responsible for recycling NAD^+ during anoxia (Sachs and Vartapetian 2007). According to the well-recognized Crawford's Metabolic Theory of the 1970s (Crawford 1977, 1978), the flooding tolerance of the submerged plants (hydrophytes) is determined by biochemical and molecular characteristics of their root and seed cells: these cells do not induce ADH activity; therefore, alcoholic fermentation and accumulation of toxic alcohol are substantially inhibited in the cells when the plants are faced with oxygen deficiency. Mesophytes, on the other hand, induce ADH activity with consequent alcoholic fermentation and toxic alcohol accumulation in the cells.

Why are the animal cells generally very sensitive to anoxia? It is because animals, on account of the lack of pyruvate decarboxylase, solely rely on lactic-acid fermentation that catalyzes a necessary intermediate step in ethanol fermentation.

Plants flourishing in wetland ecosystems are extremely tolerant to hypoxic as well as anoxic conditions. Specific vegetation is adapted to wetland environments, and this vegetation enhances the health of these ecosystems. Anaerobic microorganisms perpetuate in water-saturated soils under anaerobic conditions prevailing in the wetlands. Depleted of molecular oxygen, when waterlogged soils go completely anoxic, the role of the anaerobic microorganisms (anaerobes) in the wetlands becomes of significance in plant survival and growth. Reduction of NO_3^- to NO_2^- or N_2O and N_2 happens to the sources of energy for the anaerobes. N_2 and N_2O exit into the atmosphere during the process of denitrification. Under more severe reducing conditions, anaerobes induce Fe_3^+ into Fe_2^+. When the wetland soils continue to be anoxic for longer periods, they become toxic due to higher solubility of Fe^{2+}. Some anaerobes can reduce SO_2 into H_2S. H_2S is a respiratory poison and, thus, adds to the toxicity of wetland environments.

If rich in organic matter, the waterlogged and anoxic soils help proliferate populations of anaerobes that produce acetic acid and propionic acid. The unpleasant odor emanating from wetland areas is attributable to these acids, along with reduced sulphur compounds. All these microbial metabolites in an anaerobic environment are toxic to the plants.

ROOT DAMAGE IN ANOXIC ENVIRONMENTS

Critical oxygen pressure (COP) is the pressure of oxygen at which the respiration rate is first slowed down due to oxygen deficiency. Root degradation following a slowed-down respiration rate begins even before the complete depletion of oxygen in water-covered soils. COP is used as an indicator of the same. For instance, when growing in a nutrient solution being continuously stirred at 25°C, the maize root tips have a COP value of 20 kPa, which is at par with the O_2 concentration in ambient air (Taiz and Zeiger 2002). The rate of O_2 consumption by the roots hardly keeps pace with the O_2 supplies from dissolved oxygen. However, compared to the cells of a mammalian, the root-tips cells are metabolically more active with respect to respiration rates and ATP turnover.

When O_2 concentration goes down and becomes lower than COP, the center of the roots attain hypoxic and anoxic levels. Older roots are more prone to damage because of their lower respiration rates and consequent lower COP values. In cool climates too, the COP values are lower with the lower respiration rates. If the cells are tightly packed in a tissue, there will be minimum intercellular spaces and they (or the whole tissue) will be more prone to damage. A large number of intercellular spaces, allowing quick O_2 diffusion, as in large and bulky fruits, do not allow anaerobic conditions to prevail.

The interesting mechanism that takes place under anoxic conditions is the cessation of electron transport, oxidative phosphorylation, and the tricarboxylic cycle. In the absence of O_2, ATP is produced only by fermentation. With insufficient O_2 supplies, the plant roots first start to ferment pyruvate, emanating from glycolysis, to lactate. Lactate dehydrogenase is the enzyme involved in this conversion. However, at an acidic pH, lactate dehydrogenase is inhibited and pyruvate decarboxylase is activated. Under such conditions, only 2 mol of ATP are produced from 1 mol of hexose (glucose), whereas in the case of aerobic respiration, 1 mol of hexose yields as many as 36 mol of ATP. Thus, an adverse effect on root metabolism arises, particularly from the lack of ATPs to carry on metabolism (Figure 8.2).

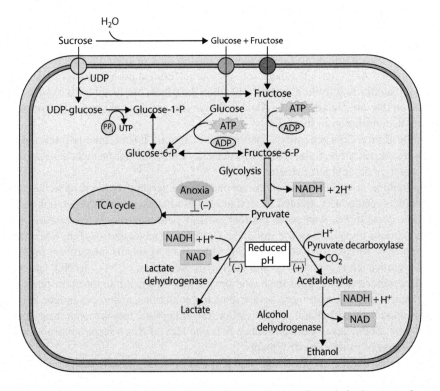

FIGURE 8.2 Alteration in the biochemical pathways during episodes of anoxia in the roots of water-logged plants. (From Taiz, L., and Zeiger, E., *Plant Physiology*, Sinauer Associates, Sunderland, 690 p.)

In healthy plant cells, the pH of the vacuolar content is more acidic than that of the cytoplasm, pH values being 5.8 and 7.4, respectively. However, an environment of extreme O_2 deficiency causes a gradual leakage of protons from vacuole to cytoplasm. This process results in lactic-acid fermentation. The changes in pH values, or cytosolic acidosis, are associated with the onset of cell death.

The lack of ATP slows down the active transport of H^+ into the vacuole by tonoplast ATPase. Loss in ATPase activity does not permit maintenance of a normal pH gradient between the cytosol and vacuole. It is the cytosolic acidosis that irreversibly disrupts the metabolism in the cytoplasm of the cells. This process can also be of use in distinguishing flood-prone plants from flood-tolerant ones.

EFFECT OF ANOXIA-DAMAGED ROOTS ON SHOOTS

In the anoxic environment, plant roots become deprived of nutrients, and their energy-production system is severely disrupted. The roots of such plants cannot absorb sufficient nutrients, and therefore, they cannot efficiently translocate them to plant shoots. Premature senescence of old leaves takes place so that phloem can transfer nutrients to the new leaves. Water potential in the leaves decreases, which is attributable to lower root permeability to water under anoxic conditions.

In some species, notably tomato and pea, stomata closure is induced during flooding of the plants. Under hypoxic conditions, the abscisic acid (ABA) production mechanism is induced, and its movement to leaves increases. Additional ABA production by the older leaves is attributable to stomata closure in the plants.

Oxygen shortage accelerates the production of 1-aminocyclopropane-1-carboxylic acid (ACC), an ethylene precursor. ACC in the tomato travels to the plant shoot via the xylem. Upon coming in contact with oxygen, the shoot ACC is converted to ethylene by ACC oxidase. High ethylene levels in the adaxial surface of the leaf petiole with ethylene-sensitive cells in sunflowers and tomatoes lead to rapid expansion of the cells, subsequently resulting in epinasty, the downward growth of the leaves (Taiz and Zeiger 2002). Loss of turgor does not take place during the episode of epinasty.

SYNTHESIS OF ANAEROBIC STRESS PROTEINS

Sachs and Ho (1986) observed that when maize roots were deprived of oxygen, protein synthesis ceased except for the continued production of about 20 polypeptides. Most of these proteins, referred to as the anaerobic stress proteins, have been identified as enzymes involved in glycolysis and fermentation.

Intracellular Ca^{2+} levels are elevated when a plant suffers from oxygen deficiency. This is how the mechanisms of reduced oxygen levels are sensed, though it is not completely understood. The elevated intercellular Ca^{2+} signal is involved in the signal transduction of anoxia. This acts as a signal, resulting in the increased mRNA levels of ADH and sucrose synthase in the maize cells (Taiz and Zeiger 2002).

Sachs et al. (1996) also found that certain chemicals that block a rise in intracellular Ca^{2+} concentrations in maize seedlings under anaerobic environment also prevent expression of the genes for ADH and sucrose synthase. Accumulation of mRNAs of the anaerobic stress genes takes place as a result of the changes in the transcription rates of these genes. Some type of translational control of anaerobic stress genes is occurring (Taiz and Zeiger 2002). The efficiency of translation of mRNAs for stress-regulated genes (e.g., ADH) in hypoxic environments is higher than that of non-anaerobic stress-regulated ones.

SALINITY STRESS

Soils throughout the globe are not uniform in their physical, chemical, and biological characteristics. There exists enormous diversity as well as several anomalies among the world's soils. Mineral composition of the soils not conducive to a plant's normal growth and reproductive functions includes high concentrations of salts (e.g., NaCl), toxic ions (Pb and As), and low concentrations of essential nutrients (N, P, K, Ca, Mg, etc.). Plants growing on such soils often face stress conditions.

Evaporation of water takes place at all temperatures, and its rate increases with increasing temperatures. This natural process leads to loss of soil water, which in turn, results in increased solute concentration. Transpiration also adds to the loss of water, leading to further increase in solute concentrations in the soils. The water lost through evaporation and transpiration is in its pure form. Solute concentration is also multiplied if crops are irrigated with water with high solute concentration. Accumulation of salts in the soils in this way goes on increasing if there is no system to flush out excessive salts into a drainage system. Under these conditions, the salt concentration in the soils reaches a level where it becomes injurious to plants, especially the salt-sensitive species. Thus, a salinity problem arises on account of both, natural processes and human activities.

Often associated with high concentrations of NaCl, the saline soils in some areas have high concentrations of Ca^{2+}, Mg^{2+}, and SO_4^-. Soils in which Na^+ occupies $\geq 10\%$ of the cation exchange capacity are the sodic soils. Such soils injure plants as well as decrease porosity and water permeability and, thus, degrade the soil structure.

Pure water in nature is very rare. Being a universal solvent, it carries certain amounts of salts dissolved in it. Salinity in soil water or irrigation water is measured in terms of electrical conductivity or osmotic potential. Conductivity of water is owing to the presence of ions in it. Pure water is a very poor conductor of electric current. Higher electrical conductivity and higher osmotic pressure (or lower osmotic potential) are indicative of higher salt concentrations of water.

EFFECT ON PHOTOSYNTHESIS AND GROWTH

The plants, on the basis of their varying saline concentrations, can be categorized as halophytes and non-halophytes. The halophytes are the natives to a saline environment, well-adapted to carry on their life cycles in these environments. The non-halophytes, also often referred to as glycophytes, are unable to tolerate saline concentrations to the same extent as the halophytes do. The glycophytes, literally meaning "sweet plants," can grow well up to a threshold salt concentration. Above this threshold, they suffer from leaf discoloration, stunted growth, and loss in biomass (dry weight).

Some of the examples of salt-tolerant and salt-sensitive plants are given in Table 8.1. Greenway and Munns (1980) observed in their experiment that two plant species, viz., a saltbush *Atriplex nummularia* and a salt marsh plant *Suaeda maritima* showed growth stimulation even when Cl⁻ concentrations were raised many times over and above the lethal level for sensitive species.

Discoloration of leaf and decreased absorption of water by plants severely affect photosynthesis and consequent performance of the plants, particularly of salt-sensitive plants. High concentrations of Na^+ and/or Cl⁻ accumulated in chloroplasts also lead to inhibition of photosynthesis. Photosynthetic electron transport is relatively insensitive to salts. As a consequence, photophosphorylation or carbon metabolism is more likely to be affected in the plants growing in saline environments.

TABLE 8.1
Some Salt-Sensitive, Moderately Salt-Tolerant, and Highly Salt-Tolerant Species

Species Response to Salt	Plant Species
Highly sensitive to salt	Citrus, pecan, maize, bean, onion, lettuce
Moderately tolerant to salt	Barley, cotton
Highly tolerant to salt	Date palms, sugar beet, *Suaeda maritime*, *Atriplex nummularia*

Osmotic and Specific Ion Effects

Osmotic potential in the root zone of the soil becomes more negative, and water potential goes down considerably when solutes are dissolved in water. Soil environment with higher solute concentrations thus affects the overall water balance of the plants. In order to strike a water balance, the plants need to further decrease water potential in their leaves. A lower leaf potential than that of the shoot, root, and soil would ensure a continuous flow of water within the plant down the concentration gradient.

The main difference between the environments with low water potential created by salinity versus soil desiccation is the total amount of water available. Plants can obtain a finite amount of water from the soil profile during the desiccation of the soil, causing an ever-decreasing water potential. Thus, a saline environment makes available quite large amounts of water at a constant and low water potential (Taiz and Zeiger 2002). More negative water potential in the soils, however, causes hindrances in availing the water to the plants as per their physiological requirements.

When growing in saline soils a number of plant species adjust osmotically, which is a case of phytosociological importance. While generating a lower water potential, loss of turgor is prevented in such an adjustment. However, the growth of the plants growing in saline environments followed by such adjustments is very slow (Taiz and Zeiger 2002).

Concentrations of ions in the saline soils up to an injurious level, especially of Na^+, Cl^-, and SO_4^{2-}, when accumulated in the cells, cause specific ion-toxicity effects.

Plants' Strategies of Saline Stress Management

In a saline environment, plants, in order to survive, can make morphological adjustments. For example, they may decrease leaf area and/or may rid their crowns of leaves through leaf abscission.

In salt-sensitive plant species' resistance to moderate levels of soil salinity, roots play crucial roles in preventing potentially harmful ions from reaching the shoots. It is a strategy of the plants to minimize salt injury by excluding salt from meristems, particularly in the shoots and leaves, which serve as sources of photosynthesis.

The Casparian strips help impose restrictions on the movement of ions into the xylem. If the Casparian strips are to be bypassed, ions will move from the apoplast to the symplastic pathway across all membranes. The salt-resistant plants derive an advantage from this transition by using these mechanisms for partially excluding toxic ions.

Saline environments are conducive for some species, for example, the mangroves. The mangrove plants have the ability to acquire water from the low-water-potential environments (saline water) by making osmotic adjustments. Plants in the saline environments can respond to osmotic stress by adjusting their water potential. They do so by decreasing their solute potential. This process is carried out through two intracellular functions, viz., (1) ion accumulation in the cell vacuoles, and (2) synthesis of compatible solutes in the cytosol (e.g., sucrose, sorbitol, mannitol, pinitol, glycine betaine, proline).

Halophytes are more capable of accumulating ions in their cellular vacuoles and thus adjusting physiological functions, enabling the plants to perform well in moderate saline concentrations. The glycophytes' capacity of accumulating ions in the vacuoles, however, is generally poor and so are the adjustments they make to survive in saline environments.

Some salt-resistant species, for example, *Tamarix* sp. and *Atriplex* sp., instead of excluding ions at the root, transport them to salt glands they have at the surface of the leaves. In these specialized glands, salts are crystallized and become harmless to the plants.

As Na^+ enters roots passively, the root cells need to invest energy to extrude the ions actively back to the soil solution and get rid of an excess of the ions. Cl^-, on the other hand, is excluded by a negative electric potential across the cell membrane. Low permeability of the root cell membrane to Cl^- is also attributable to keeping the Cl^- toxicity at a minimum.

HEAT STRESS

Heat, measured in terms of temperature, is necessary for the biochemical reactions and metabolism in organisms going on. Life in the biosphere is sustained with a certain amount of heat, or, in other words, within a range of temperatures. There could be variations in an appropriate temperature range among individual species. The range of temperatures required for plants is wider than for animals. Most of the higher plants would not survive if the temperature range toward the higher side exceeded 45°C for an extended period. Single-celled eukaryotes can survive and complete their life cycles at temperatures above 50°C, while prokaryotes can thrive at temperatures above 60°C.

Survival of organisms depends on their water requirements for maintaining physiological reactions and growth pattern. Actively growing cells or tissues, seedlings, and vegetation requiring high water quantities for growth and reproduction cannot survive if exposed to high temperatures for prolonged duration. Dehydrated tissues, such as dormant seeds and pollen grains, can withstand higher temperatures. For example, pollen grains of some plant species can tolerate and survive at 70°C, and dry dormant seeds can survive at 120°C.

WATER DEFICIT AND HEAT STRESS

Plants' access to water can cope with adverse effects of high temperatures. At high temperatures, however, water also becomes a limiting factor in terrestrial environments, which is due to an increase in evaporation rates. Thus, there is an interrelationship between water and temperature stress. Evaporative cooling in C_3 and C_4 plants with access to abundant water helps them maintain survival and functioning below 45°C. Evaporative cooling decreases if water becomes a limiting factor and, under such conditions, tissue temperature increases. If water scarcity endures, tissue temperature above 45°C becomes lethal for the plants.

C_3 and C_4 plants have to depend on transpirational cooling. The leaf temperature of the plants can rise 4°C to 5°C above ambient atmospheric temperature at midday during intense solar radiation when there is stomata closure due to soil-water deficit. This situation can also arise when high humidity diminishes water evaporation from the leaves.

The case of the crassulacean acid metabolism (CAM) plants is different from that of C_3 and C_4. CAM plants have their stomata closed during the daytime. So, they cannot maintain their temperature through evaporative cooling. They are capable of withstanding temperatures as high as 60°C–65°C. These plants lose heat only through reemission of the infrared (long wavelength) radiation from incidence radiation, or through conduction and convection.

EFFECT ON PHOTOSYNTHESIS AND RESPIRATION

Heat stress inhibits both photosynthesis and respiration. However, the rates of photosynthesis are affected prior to those of respiration. This can be elaborated in terms of the temperature compensation point, that is, the temperature at which the amount of CO_2 fixed by means of photosynthesis equals to that released by means of respiration, in a given time interval. When the temperatures rise above the compensation point, the carbon used by respiration is not replaced by photosynthesis. As a consequence, the carbohydrate reserves are exhausted at a faster pace. Heat stress, thus, leads to a decline in sweetness in fruits and vegetables.

In the same plant, the light leaves (leaves exposed to the sun) are affected prior to shade leaves. Therefore, the temperature compensation point is always higher for the light leaves than for the shade leaves. C_3 plants are greater sufferers from detrimental effects of increased respiration relative to photosynthesis than the C_4 and CAM plants. It is because higher temperatures lead to an increase in both dark respiration and photorespiration in C_3 plants.

EFFECT ON CELL MEMBRANE

High temperatures prove deleterious for cell membrane. A rapid increase in the fluidity of membrane lipids is an obvious result of the tissues' exposure to high temperatures. Disruption in the membrane-associated physiological functions is a natural consequence of heat stress.

Exposure of plant tissues to high temperatures for a longer period leads to modification in membrane composition and structure. This occurs due to slackness in the strength of hydrogen bonds and electrostatic interactions between proteins' polar groups with the aqueous phase of the membrane.

Leakage of ions could be the consequences of such temperature-induced alterations in the cell membrane. Since photosynthesis and cellular respiration depend on the activity of membrane-associated electron carriers and enzymes, these vital processes also become inhibited as a result of cellular disruptions due to heat stress.

ADAPTATION MECHANISMS AGAINST HEAT STRESS

In warm climates, plants are adapted to considerably decrease absorption of excessive solar radiation and, in this way, avoid excessive heating of their leaves. Plants in warm and somewhat inhospitable environments assume these adaptation mechanisms through special leaf structures. Leaf hairs, leaf waxes, and vertical leaf orientation aid in increasing reflection of light, thereby decreasing absorption of the radiation.

In oak (*Quercus* spp.) in the Himalayan environments, hairless leaves in the winter turn into white pubescent leaves in the summer. Bearing such dimorphic leaves is yet another example of an adaptation mechanism to avoid leaf heating by enhancing reflection of light.

High-temperature stress tolerance can be induced by short-term elevated temperatures known as acquired thermotolerance (Kotak et al. 2007). Some plant species bear highly dissected leaves to reduce the boundary thickness layer and thereby maximize heat loss through convection and conduction.

Excessive heat alters the structure and functions of many proteins that serve as enzymes in a plant. This occurs due to unfolding and misfolding of the proteins. Aggregation and precipitation of the misfolded and unfolded proteins leads to serious problems in plant cells. When the temperature rises to the tune of 5°C to 10°C, plants develop a protection mechanism by producing a unique set of proteins referred to as heat-shock proteins (HSPs). HSPs were first discovered in 1962 in *Drosophila melanogaster* (fruit fly) and later on in mammals, humans, plants, fungi, and microorganisms. Seigneuric et al. (2011) have found out that HSPs are danger signals for cancer detection.

HSPs protect the plants against heat stress, acting as molecular chaperones. The molecular chaperones help the misfolded and aggregate proteins to undergo proper folding and prevent misfolding and unwanted protein aggregation. They also help in degradation of damaged or denatured proteins. They also play a role in assisting the repair of denatured proteins as well as in promoting their degradation following heat stress or injury. This adaptation mechanism helps the plant cells function normally despite an increase in temperature up to the level of heat stress. HSPs, thus, serve as crucial molecular players in the cellular-stress response.

In response to an increase in temperatures, plants make HSPs of different sizes, with the molecular mass ranging from 15 to 115 kilodaltons (kDa). They are grouped into five classes based on their sizes (smHSP, HSP60, HSP70, HSP90, and HSP100 against sizes of 15–30, 57–60, 69–71, 80–94, and 100–114 kDa, respectively) (Taiz and Zeiger 2002). Members of HSP60, HSP70, HSP90, and HSP100 groups act as molecular chaperones involving ATP-dependent stabilization as well as folding of proteins and the assembly of oligomeric proteins. These are localized in different parts of a cell, especially the cytosol, chloroplasts, endoplasmic reticulum, mitochondria, and nucleus.

Improved thermal tolerance up to lethal temperatures has been induced in the cells induced to synthesize HSPs. There are yet HSPs not unique to high temperatures that are induced by widely different environmental stresses or under certain conditions, such as ABA treatment, salinity, and water deficit.

LOW-TEMPERATURE STRESS

Low atmospheric temperatures are not healthy for plants. Prolonged exposure to cold induces adverse changes in physiology, influencing the overall performance of the plants. Subtropical and tropical plants are especially prone to cold stress. The effect of low temperatures on plants is observed through chilling stress and freezing stress. The plants suffer from chilling stress when they are exposed to low temperatures but above 0°C (i.e., above the freezing point). Temperatures below 0°C put the plants under freezing stress. Corn, rice, cotton, sweet potato, and *Phaseolus* bean are the major crops susceptible to chilling.

In the tropics and subtropics, if warm temperatures ranging from 25°C to 35°C are cooled to 10°C–15°C, plants suffer from chilling injury, a term suggested by German plant physiologist Hans Molisch as early as 1897. Phenotypic symptoms of chilling injury include poor germination, stunted growth, chlorosis (yellowing of leaves), and reduced expansion of leaves (Yadav 2010). If the roots are affected by chilling injury, wilting of the plant takes place.

CHILLING EFFECTS ON CELL MEMBRANE

Among the biomolecules, lipids are the first to be affected by the chilling effect. Since the cell membrane is made up of a lipid bilayer interspersed with proteins and sterols, it is to be the first anatomical structure to be affected by the chilling effect. Physical properties of the lipids influence the integral membrane proteins, including the enzyme H^+-ATPase. As a result, membrane properties are altered, adversely influencing ion transport and enzyme transport and, consequently, the whole metabolism.

We know that lipids comprising saturated fatty acids and those with *trans*-monounsaturated fatty acids undergo solidification well above 0°C, that is, at higher temperatures in comparison to unsaturated fatty acids. Thus, cell membranes with a greater proportion of saturated fatty acids and *trans*-monounsaturated fatty acids in their lipid bilayer in the chilling-sensitive plant species undergo solidification into a semi-crystalline state at relatively higher (above 0°C) temperatures. The chilling-resistant plants (e.g., cauliflower, peas, and turnip) have a greater proportion of unsaturated fatty acids that tend to maintain their fluidity at much lower temperatures.

Maintenance fluidity attributable to the lipid bilayer in the cell membrane is vital for normal functioning by a cell. With the semi-crystalline state, the lipid bilayer in the cell membrane loses its fluidity. As a result, there is H^+-ATPase activity leading to deviation of the protein component of the membrane from maintaining normal functions, such as solute transport across the cell, energy transduction, and all the cellular functions dependent on enzymes.

The leaves of the plants injured by chilling show reduced rates of respiration, photosynthesis, and carbohydrate translocation. Inhibition of protein synthesis and a faster rate of protein degradation also take place in the plants suffering from chilling injury. These responses to chilling-induced injuries are primarily owing to a loss of membrane function. Inhibited rates of photosynthesis reveal that plants' exposures to chilling temperatures damage chloroplasts, and inhibition of respiration reflects damage to the mitochondria.

A list of phenotypic, anatomical, and physiological/metabolic symptoms in a response to chilling stress in plants is presented in Table 8.2.

Cold stress in plants reaches its climax when the atmospheric temperature falls down below 0°C. It occurs due to the freezing of soil water, which happens at −2°C, and also due to the freezing of cellular water. Freezing of soil water in extreme cold environments leads to blockage of water movement from the soil to the plants through the roots. At prolonged freezing temperatures, the water content inside the plants also freezes, leading to the disruption of structure and functioning of plant tissues. Ice formation inside the plants first takes place in the cell wall and intracellular spaces. When ice crystals grow, they cause punctures into the cytoplasm that damages cells and tissues. Certain large polysaccharides and proteins function as nucleating agents in the process

TABLE 8.2

Emerging Symptoms in Plants in Response to Chilling Stress

Phenotypic	Anatomical	Physiological/Metabolic
• Stunted plant growth • Cracking, splitting and dieback of stem • Abnormal curling, lobbing, and crinkling of leaves • Surface lesions of leaves and fruits • Soggy foliage (as soaked in water) • Cold hardening, altering behavior of stomata	• Changes in cell membrane structure and composition • Vascular browning • Swelling of plastid and mitochondrial membrane, dictyosomes, and chloroplast thylakoids • Plasmalemma pressed against the tonoplast and deleted into vacuoles as sac-like intrusions • Increase in plasmalemma permeability • Mitochondria with reduced crystae and transparent matrix • Extensive dilation and vesiculation of smooth ER cisternae • Accumulation of cryoprotective substances in dilated vesicular ER cisternae • Complete disappearance of rough ER with loss of ribosomes from the membrane • Transformation of rough ER into vacuolated smooth ER • Formulations of crystalline deposits in root cells as well as epidermal, vascular, and mesophyll cells of the leaves, which eventually lead to tonoplast disruption • Accumulation of lipid bodies in cytoplasm in close association with plasmalemma • Grana disintegration and increase in size and number of plastoglobules • Browning of skin and degradation of pulp tissue in banana • Blackheart of pineapple • Sunken pits in cucumber • Decrease in the size and number of starch grains	• Increased susceptibility to decay • Loss of vigor • Normal fruit-ripening failure • Loss of membrane function • Damage of chloroplast and mitochondria • Irreversible tonoplast injury • Leakage of organic and inorganic substances • Plasmolysis • Inhibited photosynthesis and respiration • Faster rate of protein degradation • Production of abnormal metabolites due to anaerobic conditions • Decreased CO_2 exchange • Disruption in amylolytic activity (conversion of starch to sugars) • Higher ratio of unsaturated to saturated fatty acids in cold-tolerant plants

of intracellular ice formation. The plasmalemma is the main site of freezing injury. Protoplasmic swelling and alteration in permeability to K^+ ions result in leakage of ions from plant tissues. After a gap of several days of freezing, a plant is left with a weakened root system and/or with split bark on stem and branches.

TOLERANCE TO COLD STRESS

When exposed to extreme temperatures deviating to intolerable limits of a natural temperature range, a plant's cellular, metabolic, molecular, and physiological dysfunctions reach a point where it becomes difficult for the plant to survive. However, there are inbuilt mechanisms through which a plant can be able to tolerate cold and freezing stress up to a certain degree. There are specialized plants that can tolerate even incredibly extreme environmental conditions. Cold or chilling tolerance denotes the ability of a plant to tolerate low temperatures without experiencing internal injury or damage, and cold acclimation is an enhanced tolerance to the physical and physiochemical vagaries of freezing stress. Cold tolerance and cold acclimation involve three processes, viz., biochemical, molecular, and metabolic.

Metabolism of the plants exposed to temperature stress (low as well as high) involves modification in metabolism in two ways (Kubien et al. 2003): (1) the plants tend to adjust their metabolism as a result of their exposure to low or high temperature, and (2) enzymes' structures, catalytic properties, and functions are changed due to temperature stress.

The cold-stress response is perceived in plants through a signal transduction. This mechanism induces the activation of transcription factors and cold-responsive genes, which, in turn, control the damage done by cold stress and help impart tolerance to the affected plants (Figure 8.3). The plants in a mesophilic temperate climate possess inducible temperature tolerance.

Enormous changes in metabolic content are associated with the acquisition of tolerance to cold stress in the plants. These include ROS detoxification activities, RNA chaperones, and

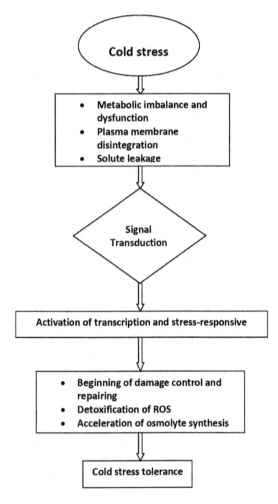

FIGURE 8.3 Cold-stress response perceived by plants through a signal transduction. (Adapted from Yadav, S.K., *Agron. Sustain. Dev.*, 30, 515–527, 2010.)

accumulation of soluble sugars and dehydrins. The temperature-stressed plants first attempt to adjust their cellular metabolism due to drastic change in temperature by altering enzyme structure, catalytic properties, and functions. Second, the plants attempt to acquire stress-tolerance by modifications in metabolism in response to temperature. The higher plants, owing to their plastic nature, are more able to react to various sorts of stresses with specific responses relating to growth.

Cold stress induces an accumulation of proline, an amino acid and an osmoprotectant. Proline, as Satoh et al. (2002) and Oono et al. (2003) have found in their experiments on *Arabidopsis*, can induce the expression of many genes with the proline-responsive element sequence ACTCAT in their promoters. Genetics plays a crucial role in determining the degree of tolerance to chilling temperatures. One important class of the proteins responsible for generating tolerance against freezing temperatures are called antifreeze proteins. These proteins are found in the fisheries thriving well in freezing environments under polar ice caps.

In natural conditions, woody plants acclimate to cold in two different stages: First, hardening is induced before the onset of the winter season (often in early autumn), that is, when the woody plants are exposed to non-freezing temperatures leading to cessation of growth. Second, direct exposure to freezing happens to be a stimulus, leading to the hardening of the plant. Fully hardened cells can tolerate plants' exposures to temperatures of the order of −50°C to −100°C.

POLLUTION STRESS ON ECOSYSTEMS

All ecosystems—in fact all organisms—in the biosphere are exposed to the influence of various environmental factors. And all organisms in most of the "modern industrially civilized" world are exposed to the phenomenal influence of pollutants. There is no dearth of our understanding about the unending trails of pollutants slowly, but steadily, poisoning our biosphere. Phenomenal impacts of many pollutants are already visible, and a large body of literature revealing how the pollution is eclipsing the Living Planet is available.

Environmental stress caused by a variety of pollutants imposes physical, chemical, and biological constraints not only on the species productivity but also on the structure and functions of an ecosystem. The environmental stress exerted by pollutants is in addition (and over and above) of what is caused by natural environmental stressors. Pollutants can be physical (intensive radiation, wildfire, high and/or low temperatures, etc.), chemical (gaseous, liquid, and solid chemicals), and biological (presence of other species/organisms) in nature. Pollution, leading to a stressful life, occurs when the pollutants are present in concentrations large enough to affect the organisms and their ecosystems.

Ecosystems are indivisible. All organisms in the community within an ecosystem are interrelated with food chains. And, after all, all the nutrients within the species that are received from the environment are ultimately recycled into their respective environmental components. Therefore, the effect of the pollutants on organisms eventually puts the whole ecosystem under stress.

Ecosystems are dynamic. Energy and nutrient flows maintain ecological integrity, homeostasis, and dynamism of an ecosystem.

Environmental factors, like excessive or insufficient radiation, temperatures, moisture—in interaction with each other and biotic factors—constitute climatic factors. Deserts and tundra are the striking examples of climatically stressed ecosystems. Tropical rainforests and other biotic communities existing in a relatively benign climatic regime, on the other hand, seldom experience climatic stress.

Radiation stress is caused due to an excessive load of ionizing energy. Long-term exposure to UV radiation, for instance, on high-altitude mountains, and due to exposure to radioactive materials, puts the organisms and their communities under severe stress.

Excessive release of heat influences and stresses the surrounding organisms and ecosystems. Wherever hot water is discharged in a water body, the whole surrounding aquatic life is affected. Thermal pollution leads to changes in the structure and composition of an aquatic community.

Excessive input of some of the nutrients, such as of nitrogen and phosphorus, in an aquatic ecosystem eventually leads to the "climax" of the water bodies' pollution emerging into a state of the eutrophication. Numerous aquatic ecosystems in our world, both lotic and lentic, are gradually dying of pollution load.

Chemicals (gaseous, liquid, and solid) toxic in nature cause widespread hazards in ecosystems, bringing the community under a variety of stresses. Gases (e.g., CO, SO_2, N_2O, and O_3), pesticides (weedicides, insecticides, fungicides, rodenticides, etc.), and heavy metals (Hg, As, Pb, etc.) acutely affect the whole community in various ways through their deleterious effect on the anatomy, biochemistry, and physiology of all the organisms exposed to chemical pollution (Negi et al. 2019). As a consequence, ecological processes have to be affected by toxic pollutants.

HEAVY-METAL STRESS

The term "heavy metals" is often used as a group name for metals and semimetals (metalloids) that are associated with contamination and potential toxicity or ecotoxicity (Duffus 2002; Negi et al. 2018b). The heavy metals represent a large group of chemical elements (>40) with relatively high density, high atomic numbers, and high atomic mass. Most of the heavy metals may be important trace elements in the nutrition of plants, animals, or humans (e.g., Zn, Cu, Mn, and Fe), while others (e.g., Pb, Cd, and Hg) are not known to have positive nutritional effects. All these metals (and metalloids), however, may have toxic effects, some of them at a very low content level (Spiegel 2002). Heavy-metal toxicity, to a great extent, depends on their chemical form, concentration, residence time, etc. (Mielke and Reagan 1988). One of the characteristics associated with these elements is that they do not decay with time. Thus, their emission into the environment—increasing worldwide due to various anthropogenic activities—is increasingly becoming a serious problem.

Rapid industrialization, coal-based power plants, petroleum refineries, mining operations, smelters, foundries, urbanization, mismanagement of wastes, indiscriminate applications of pesticides, use of wastewater for irrigation, etc. are the various causes giving rise to an accumulation of heavy metals in soils, water bodies, atmosphere, plants, and animals, contributing to incalculable stress on ecosystems. Pb, Cd, Hg, Cr, Ar, and Cu are the major heavy-metal pollutants. Heavy-metal pollution, due to ever-intensifying anthropogenic activities, has now become an issue in many areas of the world (Nagajyoti et al. 2010). Some metals are required by plants and animals in trace amounts but are detrimental and toxic for flora and fauna at higher concentrations (Alloway 1995; Parmar and Singh 2015a, 2015b; Negi et al. 2018a). Heavy-metal pollution in aquatic ecosystems is caused by means of toxicity, reduced growth, and morphological, histological, and behavioral changes among organisms.

REACTIVE OXYGEN SPECIES

Stress conditions commonly lead to the generation of ROS, such as $\cdot O_2^-$, H_2O_2 and $\cdot OH$, that bear strong oxidizing activities that can attack all types of biomolecules. ROS represent intermediates emerging during the successive reduction of O_2 to H_2O. Exposure of the plants to certain heavy-metal ions results in shifting the balance of free-radical metabolism toward an accumulation of H_2O_2. If the redox active-transition metals, such as Cu^+ and Fe^{2+}, are present, H_2O_2 can be converted to the highly reactive $\cdot OH$ molecule in a metal-catalyzed reaction via the Fenton reaction (Negi et al. 2018a). The oxidized metal ions undergo a re-reduction in a subsequent reaction with superoxide radicals ($\cdot O_2^-$). An alternative mechanism of the $\cdot OH$ formation directly from H_2O_2 and $\cdot O_2^-$ is the metal-independent Haber–Weiss reaction.

$$Fe^{3-} + \cdot O_2^- \rightarrow Fe^{2-} + O_2$$

$$Fe^{2-} + H_2O_2 \rightarrow Fe^{3+} + OH^- + \cdot OH$$

$$\cdot O_2^- + H_2O_2 \rightarrow \cdot OH + OH^- + O_2$$

INACTIVATION OF BIOMOLECULES

Heavy metals interact with biomolecules and block essential functional groups or displace essential metal ions.

PRIMARY STRESS

Heavy-metal stress results in the inhibition of photosynthesis and transpiration and also suppresses metabolic processes vital for the sustainability of life.

SECONDARY STRESS

Once the primary metabolic functions come to a standstill mode, thanks to the effects of heavy metals, the natural repercussions that take place are senescence and nutritional deficiency, eventually leading to death of the plant.

SIGNAL TRANSDUCTION

Changes in the gene expression pattern occur upon encountering heavy metals. For example, in *Arabidopsis thaliana*, mitogen-activated protein kinase (MAPK) gets charged up and leads to the activation of the phosphorylation cascade as follows:

MAPKKK (Ser-Thr protein kinase) → MAPKKs (Thr-Tyr protein kinase) → MAPKs (Ser-Thr protein kinase) → activation of substrates in various organelles

EFFECTS ON CELLULAR COMPONENTS

The lipid peroxidation chain begins taking place due to the production of the ROS species in chloroplasts, plasma membranes, mitochondria, peroxisomes, apoplasts, and lipid peroxidation, on account of which the unsaturated fatty acids are converted to saturated ones leading to wall disruption (Gill and Tuteja 2010).

The thiol group in proteins is susceptible enough to attract metal ions to attack them, rendering the proteins functionless. With the proteins becoming deconformed, DNA mutation and genetic defects become obvious. When metals penetrate the cell wall, they can damage free carbohydrates and cell-wall polysaccharides (Moller et al. 2007).

HEAVY-METAL DETOXIFICATION METHODS

There is a range of potential cellular mechanisms a plant employs for heavy-metal detoxification and thus for tolerance to metal stress (Hall 2002). There are different ways plants respond to heavy-metal toxicity, for example, immobilization, exclusion, chelation, and compartmentalization of the metal ions, as well as the expression of ethylene and stress proteins through more general stress-response mechanisms. An understanding of the molecular and genetic basis relating to these mechanisms will be crucial for developing plants into agents for the phytoremediation of contaminated sites (Negi et al. 2018a, 2018b).

One recurrent general mechanism for heavy-metal detoxification in plants and other organisms is the chelation of the metals by a ligand and, in some cases, the subsequent compartmentalization of the ligand–metal complex.

- Translocation-soil/water → roots → cell wall → cell membrane → cytosol → vacuole
- Sequestration → chelation
- Detoxification → volatilization and compartmentalization

The roles of several metal-binding ligands recognized in plants have been reviewed by several workers, notably Rauser (1999), Hall (2002), Parmar and Singh (2015a, 2016), Negi and Singh (2016), Negi et al. (2018a, 2018b), and Negi et al. (2019). Extracellular chelation involving organic acids (e.g., citrate and malate) is of key importance in mechanisms of aluminum tolerance. Exposure to aluminum stimulates malate efflux from root apices and is correlated with aluminum tolerance in wheat as suggested by Delhaize and Ryan (1995). Some aluminum-resistant mutants of *Arabidopsis* also have increased organic acid efflux from roots (Larsen et al. 1998). Some amino acids, especially His and some organic acids, also play roles in the chelation of metal ions in xylem sap as well as within cells (Kramer et al. 1996; Rauser 1999). Peptide ligands include the metallothioneins (MTs), small gene-encoded, Cys-rich polypeptides. The phytochelatins (PCs), on the other hand, are enzymatically synthesized Cys-rich peptides. Our understanding about PC biosynthesis and function in recent years is predominantly based on molecular-genetic approaches using model organisms (Negi et al. 2018a).

The historical context of the MTs and PCs identification in plants is worth discussing. MTs were the first to be identified as Cd-binding proteins in mammalian tissues. MTs were assumed to be the early reports of metal-binding proteins. After the elucidation of PC structures, it was found that these peptides are distributed widely in the plant kingdom, and the PCs were proposed to be the functional equivalent of MTs (Grill et al. 1987). Several examples of MT-like genes, and in some cases MT proteins, have subsequently been isolated from a variety of plant species. Most likely, the two play relatively independent functions in metal detoxification and/or metabolism. In an animal species, PCs are not reported, a fact that supports the notion that MTs may well perform some of the functions in animals normally contributed by PCs in plants. The PC synthase gene from plants and the consequent identification of similar genes in animal species suggest that both of these mechanisms contribute to metal detoxification and/or metabolism in animal species.

The classification of plants on the basis of their tendency to accumulate metals (Raskin et al. 1994) are as follows:

1. Cu/Co
2. Zn/Cd/Pb
3. Ni

Class of Histone methyltransferases (HMT) proteins:

1. Cpx-type heavy-metal ATPases—overall metal homeostasis
2. Nram—natural-resistance-associated macrophage protein family
3. CDF—cation diffusion facilitator
4. ZIP—zinc-iron permease family
5. ITR 1—subfamily of ZIP

Volatilization is yet another method of detoxification. It involves the conversion of metal ions into the vapor state. Example: $Hg^2 \rightarrow Hg^0$ Mer A and Mer B

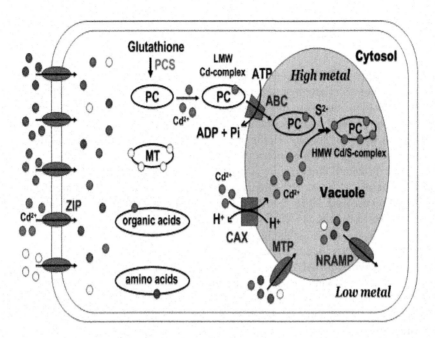

FIGURE 8.4 Vacuolar sequestration of heavy metals in the plant cell. (From Yang, Z. and Chu, C., Towards understanding plant response to heavy metal stress, doi:10.5772/24204, 2011.)

VACUOLAR COMPARTMENTALIZATION

A vacuole in a plant cell is considered to be a space for the storage of heavy metals. Vacuolar compartmentalization, therefore, is quite an effective way of controlling the concentration and distribution of metal concentration within plants.

In the vacuoles, heavy metals are arrested into a space and are rendered unavailable for other parts of the plants, helping to avoid the toxic effects of the metals. The following are the steps of vacuolar compartmentalization (Figure 8.4) (Negi et al. 2018a):

1. The ZIP transporter carries the metal ions inside the cell wall.
2. The glutathione with the help of plastocyanin synthase generates PC that binds the metal ions.
3. The low molecular weight ion-PC (phytochelatins) complex moves inside the vacuole by the ABC transporter. ABC transporters constitute one of the largest protein families in all organisms and are driven by ATP hydrolysis (Figure 8.4). They serve as exporters as well as importers. The plant genome has been found encoding for more than 100 ABC transporters, far more than in other organisms (Kang et al. 2011).
4. The low molecular weight (LMW) ion complex combines with more metal ions transported inside by the cation/proton exchangers (CAX) to form the high molecular weight (HMW) ion complex.
5. In this way, the metal ions are concentrated inside the cells' vacuoles, which later on can be excreted out.

POLLUTION STRESS IMPACTS ON ECOLOGICAL PROCESSES

An unprecedented and rapid change in environmental conditions is gradually—perhaps irreversibly—overriding the adaptive potential of the plants. Air pollution, water pollution, soil pollution, and pollution in every walk of life with the entire environmental mess gradually and dynamically culminating in climate change originating from anthropogenic activities are all injecting stress, not only into individual plants but also in the ecosystems they are integral parts of.

Diverse interspecies and intraspecies interactions occur in an ecosystem, which tend to induce the conditions of biological stress in an ecosystem. An increase in a species' population due to favorable abiotic or biotic factors or due to unwarranted human intervention catalyze the conditions of intensive competition among individuals of the populations. The same situation prevails among the individuals of different species. Herbivory, carnivory, parasitism, pests, and diseases are the other agents proliferating stress conditions in an ecosystem.

Various types of natural stresses prevailing in an ecosystem are part of the natural ecological processes. For instance, herbivory, carnivory, parasitism, pests, and diseases are the elements that help in maintaining ecological balance. However, when abiotic of biotic factors are in excess or in deficiency, the stresses caused can affect the very ecological balance of an ecosystem. Among the biological agents causing stress are the human beings who can, and they often do, modify the ecosystems and thus change the definition of ecological stress.

The prevailing environmental stresses multiply the deleterious effects of pollutants and vice versa. Liess et al. (2016) found that in the presence of non-chemical environmental stressors, an individual's sensitivity to toxic pollutants (trace metals and pesticides) multiplies by a factor of up to 100. To predict environmental stress and environmental pollution relationship, they have developed the "Stress Addition Model" and found that this kind of relationship contributes to the global biodiversity crises.

Long-term intensity of environmental stress coupled with the load of pollutants amidst the ongoing climate change leads to various sorts of ecological adjustments including alteration in biogeochemical cycles, decreased photosynthetic rates, declined productivity and decomposition rates, reduced species richness, elimination of vulnerable species, replacement of native species by more tolerant or alien species, and biodiversity erosion. When the intensity of various environmental stresses causes significant changes, ecological disruption leading to mortality of individuals is an inevitable consequence. Ecological disruption infesting a variety of ecosystems over a large area or in a vast geographical region over a long period of time pushes many species toward extinction. Such ecological responses to constant and intensified environmental stresses eventually lead to a situation of ecological conversion. Anthropogenic activities leading to ecosystems' modifications and further intensifying biotic stresses advance the state of "ecological conversion" toward what we can call an "ecological coup."

WATER STRESS

The biosphere is never in a static state. It is usually in a dynamic state. The living organisms in the biosphere constantly encounter abiotic stresses we have discussed so far in this chapter. The most common problem across the world that organisms frequently encounter is perhaps that of water stress. There is hardly any terrestrial ecosystem that does not experience water shortages at some point. Some plants, even in the rainforests as we previously discussed, struggle to minimize water loss by means of midday wilting. In a forest community, trees are relatively more stable and less prone to the abiotic stresses than animals. However, cultivated plants or the food-crop annuals often fall prey to water stress. As a result, crop failures, often leading to disruption in food supplies, are the common scenarios emerging each year in several parts of the world. Average yields of the major crop plants owing to stresses, according to Lisar et al. (2012), may decrease to the extent of 50% and even more. Water is central to all the physiological processes of plants. Withdrawal of water triggers the processes that retard, arrest, and/or terminate normal physiological functions of plants. There are, however, some plant species (or their genetic varieties) that are quite tolerant to drought conditions, especially at their grown-up stage, and can thrive even in inhospitable habitats facing acute water crises.

THE ETIOLOGY OF WATER STRESS

There are many reasons why plants are caught into a state of water stress. This inhibiting condition prevails due to many factors operating on an environment. There are seven major reasons, viz.

1. Geographical setting
2. Drought spell
3. Soil salinity
4. Low soil temperature
5. Limited water supply
6. High transpiration rates
7. Soil fertility management (anthropogenic factors)

Some geographical areas or agro-ecological zones witness water stress on a continuous basis. There are the regions/zones falling under hot, arid, or semiarid climatic areas characterized by low rainfall and frequent longer summer periods. Some mountainous, hilly, or upland areas with shallow soils and poor water-holding capacity face water stress conditions every now and then.

Some agro-ecological zones even in the hospitable climate may experience prolonged abnormal rainfall, leading to scarcity of water and consequent water stress for short, if not long, periods. This kind of drought spell occurs due to some erratic climate pattern, monsoon failure, etc. and can be expected anywhere in the world.

A condition often referred to as "physiological" drought prevails in saline soils. In these soils, water may be there in the soil solution, but it does not become available for the plants. Plants cannot absorb water with high-solute concentration because of considerable decline of the water potential not allowing water movement from the saline soil to the roots of the plant.

Low soil temperatures also cause a decrease in the water movement and absorption in plants, which is attributable to the increased viscosity of water, coupled with decreased permeability of the root-cell membrane. It implies that even if water availability is adequate, low soil temperature might induce water stress. The effect of low soil temperature, especially in the cold climates, is not as high as of the species growing in hot climates.

Water deficit is caused when water supply to plant roots is limiting. In this situation, plants suffer from water stress. This situation is rare in case of deep-rooted woody perennials (trees and shrubs) but can occur in case of shallow-rooted annuals, especially in annual food crops.

High rates of transpiration create a water deficit in the plants switching over to water stress. Transpiration rates occur in response to light and hot temperatures. While water stress due to high transpiration rates is quite common in hot and dry climates, it also prevails even in the humid climates, including in rainforests, at a certain point of time (e.g., at midday).

Anthropogenic factors are often the dominant ones in inducing (or avoiding) water stress in plants. Application of high doses of chemical/inorganic fertilizers results in increased water demands by the plants. If adequate water supplies are not made, the plants come under water stress. If a greater proportion of organic matter (including humus) is there, the soil will hold more amounts of water in its environment, and the plants growing in such soils will not be under stress if adverse environmental conditions do not prevail for prolonged periods.

MULTIDIMENSIONAL NATURE OF WATER STRESS

Since water is central to life processes, its scarcity to a significant extent (e.g., during prolonged drought spell) is of multiple dimension in its nature. Sustained drought conditions would ultimately lead to dehydration and death of the plants. However, many morphological, anatomical, cytological, biochemical, physiological, and molecular changes will take place before the plant dies of prolonged water stress. These have been discussed under the following sections.

Morphological, Anatomical, and Cytological Changes

Prolonged water stress has its first visible impact on the leaves of a plant. The plant leaves shrink, and their size decreases. The leaf cell wall thickens. Cutinization of the leaf surface occurs. With shrinkage in leaf size, the numbers of stomata also decrease, which is the first anatomical symptom of water stress. The other symptoms are shown in Table 8.3.

The morphological changes do not appear in the first instance. They are the biophysical dimensions of the anatomical and cytological changes taking place under the water-scarcity conditions.

Photosynthesis

Since water is an input of the photosynthetic reaction, photosynthesis has to be adversely affected when plants face a state of water deficiency. In the event of sustained water deficiency, structural and functional alterations take place, and the photosynthetic apparatus is hit directly. Higher plants suffer due to a decline in leaf-water potential and relative water content with subsequent retardation in photosynthesis. Decreased photosynthetic rates are primarily on account of impairment of photosynthetic pathways and closure of stomata. The latter might not be the primary cause in every case. For example, stomata closure is especially related with the inhibition of C_4 photosynthesis. Therefore, on the whole, metabolic impairment triggered by water deficiency, is the major cause of photosynthetic inhibition.

Although both C_3 and C_4 plants register a decrease in photosynthetic rates, C_4 plants are more susceptible to water stress than C_3 plants (Lisar et al. 2012).

A number of co-factors emerge within the plants when they are caught in prolonged water-stress conditions. These are:

1. Changes in photosynthesizing pigments—quantitative as well as qualitative;
2. Reduced CO_2 uptake on account of stomata closure;
3. Poor assimilation rates in plant leaves due to (a) reduced enzymic activity, (b) inhibition in chloroplast activity, (c) decline in photosynthetic metabolites, and (d) reduced carboxylation efficiency; and
4. Damage of photosynthetic apparatus, which is attributable to the production of ROS, for example, superoxide and hydroxyl radicals.

Water stress reduces the very basis of photosynthesis: the chlorophyll. Chlorophyll synthesis inhibition owing to water stress involves four stages, viz. (Lisar 2012):

1. Formation of 5-aminolevulinic acid, $C_5H_9NO_3$ (5ALA);
2. Condensation of 5ALA into porphobilinogen ($C_{10}H_{14}N_2O_4$) and primary tetrapyrrole, which is transformed to protochlorophyllide ($C_{35}H_{32}MgN_4O_5$), an immediate precursor of chlorophyll a;

TABLE 8.3
Morphological, Anatomical, and Cytological Responses to Prolonged Water Stress

Morphological	Anatomical	Cytological
• Decrease in leaf size	• Decrease in stomata numbers	• Thickening of cell wall
• Cutinization of leaf surface	• Underdevelopment of conductive system	• Reduced number of cells
• Increase in root:shoot ratios	• Early senescence induced	
• Decrease in root dry weight	• Submersion of stomata in xerophytes	
• Decrease in total leaf area		
• Decrease in leaf-area plasticity		

---

3. Light-dependent conversion of protochlorophyllide into chlorophyllide ($C_{35}H_{34}MgN_4O_5$); and

4. Synthesis of chlorophyll a ($C_{55}H_{72}MgN_4O_5$) and chlorophyll b ($C_{55}H_{70}MgN_4O_6$) with inclusion into developing pigment-protein complexes of photosynthetic apparatus.

Chlorophyll, in most of the cases, is more sensitive to water scarcity than carotenoids. An increase in xanthophylls pigments (e.g., zeaxanthin and antheraxanthin) in the plants suffering from water stress has been observed. The xanthophyll cycle has an inhibitory role on ROS production, and some of the pigments are part of this cycle. The xanthophylls, thus, seem to play a protective role in the plants facing water stress.

Stomata closure as a result of water deficiency creates conditions for reduced supplies of CO_2 to the plants. This leads to an increase in photorespiration, a negative implication for the plant. In the Calvin cycle, Ribulose-1,5-bisphosphate carboxylase/oxygenase (RuBisCO) functions as a carboxylase, and in photorespiration it acts as an oxygenase. Water stress rapidly brings down the RuBisCO content, which, as a consequence, results in declined activity of the enzyme. Hence, a reduction in the oxygen-free radicals' production and subsequent oxidative damage in chloroplasts takes place. Water stress also causes a considerable shrinkage in the chloroplast of the leaves. RuBisCO is further reduced due to acidification of chloroplast stroma. Both chloroplast shrinkage and RuBisCO reduction combine to adversely affect plant metabolism amidst the depressed photosynthetic machinery of the plant. Not only of RuBisCO, but also the activity of other photosynthetic enzymes (e.g., NADP-dependent glyceraldehyde phosphate dehydrogenase, PEP carboxylase, fructose-1,6-bisphosphatase, phosphoribulose kinase, sucrose phosphate synthase, and NAD-dependent malate dehydrogenase) is reduced to a varying extent due to water stress.

During the light reaction of photosynthesis, water stress causes a disruption of the cyclic and noncyclic electron transport chain. This disruption is in addition to the negative effect of water stress on dark reactions of photosynthesis. Rates of photophosphorylation, ATP synthesis, and NADP+ reduction are slowed down. Inhibition of ATPase leads to decreased ATP levels in the apparatus of photosynthesis. Although water deficiency adversely affects both PS I and PS II, the negative effect on PS I is relatively more severe.

Protein Synthesis

A drought spell leads to a decreased number of leaves. As a result, there is a reduced size (area) of the crown cover. This decrease is on account of suppressed rates of leaf-protein synthesis. A decrease in leaf numbers in C_3 plants is more drastic than in C_4 plants, obviously because the latter can withstand more water deficiency.

Water stress induces changes in plant proteins quantitatively and qualitatively. It is because water paucity alters gene expression and, thereby, synthesis of mRNA and new proteins. The proteins synthesized in response to water stress are the desiccation stress proteins responding to ABA, late embryogenesis abundant (LEA)-type proteins, cold regulation proteins, dehydrins, proteases, and detoxification or antioxidant enzymes, such as ascorbate peroxidase (APX), catalase (CAT), glutathione reductase, peroxidase, and superoxide dismutase (SOD). Protein kinases and transcription factors, the factors involved in the regulation of signal transduction and gene expression, are also synthesized under the water-stress conditions. The stress-response proteins are mostly dehydrin-like proteins that accumulate during embryo maturation and seed production, especially in higher plants (Csiszár et al. 2007; Lisar et al. 2012).

The major types of stress-induced proteins include LEA-type proteins and HSPs. These proteins play a role in protecting macromolecules like enzymes, mRNA, and lipids from dehydration.

Proline Accumulation

The most significant and immediately detectable change in the water-deprived plants is the accumulation of free proline. A few drought-resistant crop varieties accumulate more proline than the drought-susceptible ones, and a relationship between proline accumulation and yield stability index in water-stressed barley genotypes has also been experimentally shown (Kandpal et al. 1981). However, Zlatev and Stoyanov (2005) suggested that proline has no role to play in the stress-tolerance mechanism and that it is only useful as a "drought injury sensor."

Proline accumulation in drought-affected plants is usually considered as an osmoprotection agent. It also has a role to reduce the oxidative damage by decreasing and/or scavenging the free radicals. Contrary to the findings of Zlatev and Stoyanov (2005) is the conclusion derived by Vendruscolo et al. (2007) on the basis of their experiments, which states that proline plays role in the tolerance mechanism against oxidative stress and that proline accumulation during the period of water stress is the main strategy of the plants helping them avoid the detrimental impacts of water stress.

Lipids

Peroxidation of lipids is imminent due to water stress. The association between membrane lipids and proteins—lipoproteins—is severely affected. Enzyme activity and transport capacity of membranes is also affected. The fatty-acid composition of lipids is altered; for example, in chloroplasts there is an increase of fatty acids with less than 16 carbons.

Mineral Nutrition

Water stress affects mineral nutrition as well as causes a disruption in homeostasis. An acute decrease in Ca^{2+} in maize leaves on account of water stress has been observed (Lisar et al. 2012). The structural and functional integrity of the plant membrane is affected due to a fall in Ca^{2+} concentration because this nutrient has an essential role to play in maintaining this integrity. Membrane damage and disruption in ion homeostasis phenomenally cause a fall in K^+ contents when a plant is exposed to water-stress conditions. A decline in K^+ concentration further reduces plants' resistances to water stress. In-plant injury is also a notable incident faced by a plant due to stress-induced disruption in nitrogen metabolism. A fall in the level of nitrate reductase and NO_3^- uptake is also one of the consequences of water-stress conditions.

Abscisic Acid Accumulation

ABA, a plant hormone that plays a role in many plant-development processes, gets accumulated in the plants under water-stressed conditions, playing a role in the response and tolerance to dehydration. There is a marked increase in the ABC contents in the xylem in response to water deficiency, leading to an increase in ABC concentrations in plant leaves. An inhibition in PM-ATPase activity raises the cell wall pH, resulting in ABA^- formation. As ABA^- cannot penetrate the plasma membrane, it is translocated into guard cells. As a consequence, high ABA concentrations around guard cells stimulate the mechanism of stomata closure and the resulting water conservation by the plants under drought conditions.

Oxidative Stress

Drought is a type of oxidative stress enhancing the generation of active oxygen species or reactive oxygen species (ROS) at the cellular level. Water stress induces inhibition in photosynthesis by creating an imbalance between electron generation and utilization. The ROS, such as superoxide radicals ($\cdot O_2^-$), hydroxyl radicals ($\cdot OH$), singlet oxygen ($^1O_2$), and hydrogen peroxide (H_2O_2) are generated as a result of excess light energy received at the photosynthetic apparatus (Tatar and Gevrek 2008). Denaturation of structural and functional macromolecules is the well-understood result of ROS generation in plant cells. Apart from water stress and drought conditions, the oxidative stress is also generated due to high temperature, salinity, etc.

ROS damages cell membranes and macromolecules. They also affect cellular metabolism. Cellular damage under water-stress conditions is a frequent case in the plants. Lipid peroxidation and oxidation of photosynthetic pigments are the other detrimental effects of ROS.

Plants control the levels of ROS by means of different enzymatic and non-enzymatic scavenging mechanisms. SOD converts superoxide radicals to hydrogen peroxide. CAT enzymes and other enzyme systems (e.g., APX) remove cellular H_2O_2.

In response to ROS generation, thanks to drought conditions and other abiotic factors, the plant cells enhance expression of certain genes for antioxidant functions and the production of stress proteins with an up-regulation of antioxidant systems, which contribute to the enhanced scavenging capacity against ROS.

SUMMARY

Stress conditions emanate from unfavorable environmental conditions. In some ecosystems all the time and in some at certain point of time, some or all organisms experience stress conditions and tend to combat and adapt to them. In the visible range of solar spectrum, irradiance far above the light saturation point of photosynthesis often causes light stress leading to disrupting chloroplast structure, photoinhibition, and reduced photosynthetic rates. The excess of light causes direct damage of D1 protein that occurs as a result of excess light excitation at PSII reaction center and consequent inactivation of the same. Exposed to excessive solar radiation, pigments in plant leaves absorb excessive light energy, thereby producing excessive electrons far more than the availability of NADP+ to act as electron sink at PSI. Excess electrons produced by PSI lead to the production of ROS, the reactive oxygen species, especially the superoxide (O_2^{e-}). Oxidative damage of proteins, lipids, DNA, and RNA is done by the ROS, including superoxide, the low molecular weight compounds.

Plants often experience stress due to (i) high temperatures, (ii) low temperatures above melting point, and (iii) low temperatures below freezing point. Both photosynthesis and respiration are readily affected by temperature stress. Photosynthesis is more sensitive to temperature fluctuations than respiration. High temperatures inhibit rates of photosynthesis more than those of the respiration and strike an imbalance between photosynthesis and respiration, which is owing to the deleterious effects of the heat on these two vital processes. An increase in membrane lipid fluidity coupled with a decrease in hydrogen-bonding strength and knocking down of three-dimensional structures of proteins, including enzymes and, thus, leakage of ions from the cells, also occurs in an environment suffering from high temperatures.

Low temperatures cause damage to the plants in several ways, and extremely low temperatures for extended period might even be lethal to the plants. Shearing of cell organelles and cell membrane, intercellular crystallization of the water, external crystallization followed by cellular dehydration, decrease of water potential in the apoplast and increase in the symplast, and severe dehydration are the main consequences of the plants' exposure to low temperatures.

Plants occurring in wet and flooded soils frequently suffer from hypoxia and anoxia. Damage of plant roots and seeds, lower water diffusion rates, accumulation of toxic chemicals, damage of roots, etc. are the major consequences of the stress due to oxygen deficiency or anoxic conditions.

The salinity stress in plants growing in saline soils causes adverse effects on photosynthesis and plant growth, and osmotic potential. Plants can obtain only a finite amount of water from soil profile during the desiccation of the soil causing a decrease in water potential. Often associated with high concentrations of NaCl, the saline soils in some areas have high concentrations of Ca^{2+}, Mg^{2+}, and SO_4^-. Soils in which Na^+ occupies $\geq10\%$ of the cation exchange capacity are the sodic soils. Such soils injure plants as well as decrease porosity and water permeability and, thus, degrade soil structure. Saline environments are conducive for some species, for example, the mangroves. Halophytes are more capable to accumulate ions in their cellular vacuoles than glycophytes, thus adjusting physiological functions enabling the plants to perform well in moderate saline concentrations.

Heat stress leads to water deficit and inhibits photosynthesis and respiration. The rates of photosynthesis are affected prior to those of respiration. The carbohydrate reserves are exhausted at a faster pace. Heat stress, thus, leads to a decline in sweetness in fruits and vegetables. Exposure of plant tissues to high temperatures for longer period leads to modification in membrane composition and structure.

Prolonged exposure to cold induces adverse changes in physiology influencing the overall performance of the plants. In the tropics and sub-tropics if warm temperatures ranging from 25–35°C are cooled to 10–15°C, plants suffer from chilling injury. Membrane properties are altered, thereby adversely influencing ion transport and enzyme transport and, consequently, the whole metabolism. Leaves of the plants injured by chilling show reduced rates of respiration, photosynthesis, and carbohydrate translocation.

Environmental stress caused by a variety of pollutants imposes physical, chemical, and biological constraints not only on the species productivity but also on the structure and functions of an ecosystem. Stress caused due to heavy metals commonly leads to the generation of ROS, such as $^{\bullet}O_2^-$, H_2O_2, and $^{\bullet}OH$, that bears strong oxidizing activities that can attack all types of biomolecules.

Water stress is among the most common stresses the plants fall prey to in almost all ecosystems due to a variety of reasons, such as geographical setting, drought spell, soil salinity, low soil temperature, limited water supply, high transpiration rates, and poor soil fertility management. Prolonged water stress due to drought spell leads to morphological, anatomical, and cytological changes; reduced efficiency of photosynthesis; adverse effects on protein and lipid synthesis; poor mineral nutrition; abscisic acid accumulation; proline accumulation; and oxidative stress.

REFERENCES

Alloway, B. J. (ed.). 1995. *Heavy Metals in Soils.* Berlin, Germany: Springer. 368 p.
Armstrong, W. 1979. Aeration in higher plants. *Adv. Bot. Res.* 7: 225–332.
Crawford, R. M. M. 1977. Tolerance of anoxia and ethanol metabolism in germinating seeds. *New Phytol.* 79: 511–517.
Crawford, R. M. M. 1978. Metabolism adaptation to anoxia. In: Hook, D. D. and Crawford, R. M. M. (eds.) *Plant Life in Anaerobic Environments.* Ann Arbor, MI: Ann Arbor Science, pp. 119–136.
Csiszár, J., Lantos, E., Tari, I., Madoşă, E., Wodala, B., Vashegyi, Á., Horváth, F. et al. 2007. Antioxidant enzyme activities in *Allium* species and their cultivars under water stress. *Plant Soil Environ.* 53 (12): 517–523.
Delhaize, E. and Ryan, P. R. 1995. Aluminium toxicity and resistance in plants. *Plant Physiol.* 107: 315–321.
Duffus, J. H. 2002. "Heavy metal"- A meaningless term? *Pure Appl. Chem.* 74 (5): 793–807.
Gill, S. S. and Tuteja, N. 2010. Reactive oxygen species and antioxidant machinery in abiotic stress tolerance in crop plants. *Plant Physiol. Biochem.* 48: 909–930.
Greenway, H. and Munns, R. 1980. Mechanisms of salt tolerance in nonhalophytes. *Annu. Rev. Plant Physiol. Plant Mol. Biol.* 31: 149–190.
Grill, E., Winnacker, E. L. and Zenk, M. H. 1987. Phytochelatins, a class of heavy-metal-binding peptides from plants are functionally analogous to metallothioneins. *Proc. Natl. Acad. Sci. USA* 84: 439–443.
Grime, J. P. 1977. Evidence for the existence of three primary strategies in plants and its relevance to ecological and evolutionary theory. *Am. Nat.* 111: 1169–1194.
Hageman, R. H. and Flesher, D. 1960. Nitrate reductase activity in corn seedlings as affected by light and nitrate content of nutrient media. *Plant Physiol.* doi: https://doi.org/10.1104/pp.35.5.700.
Hall, J. L. 2002. Cellular mechanisms for heavy metal detoxification and tolerance. *J. Exp. Bot.* 53: 1–11.
Jackson, M. B. and Ram, P. C. 2003. Physiological and molecular basis of susceptibility and tolerance of rice plants to complete submergence. *Ann. Bot.* 91: 227–241.
Kandpal, R. P., Vaidyanathan, C. S., Udaya Kumar, M., Krishna Shastry, K. S. and Appaji Rao, N. 1981. Alterations in the activities of the enzymes of proline metabolism in Ragi (*Eleusine coracana*) leaves during water stress. *J. Biosci.* 3 (4): 361–370.
Kang, J., Park, J., Choi, H., Burla, B., Kretzschmar, T., Lee, Y. and Martinoia, E. 2011. Plant ABC transporters. In: *The Arabidopsis Book*, vol. 9. Rockville, MD: American Society of Biologist.
Kotak, S., Larkindale, J., Lee, U., von Koskull-Doring, P., Vierling, E. and Skarf, K. D. 2007. Complexity of the heat stress response in plants. *Plant Biol.* 10: 310–316.
Kramer, U., Cotter-Howells, J. D., Chamock, J. M., Baker, A. J. M. and Smith, J. A. C. 1996. Free histidine as a metal chelator in plants that accumulate nickel. *Nature* 379: 635–638.

Kubien, D. S., von Caemmerer, S., Furbank, R. T. and Sage, R. F. 2003. C4 photosynthesis at low temperature: A study using transgenic plants with reduced amounts of rubisco. *Plant Physiol.* 132: 1577–1585.

Larsen, P. B., Degenhardt, J., Stenzler, L. M., Howell, S. H. and Kochian, L. V. 1998. Aluminium-resistant Arabidopsis mutant that exhibit altered patterns of aluminium accumulation and organic acid release from roots. *Plant Physiol.* 117: 9–18.

Liess, M., Foit, K., Knillmann, S., Schäfer, R. B. and Liess, H.-D. 2016. Predicting the synergy of multiple stress effects. *Sci. Rep.* 6: 1–8.

Lisar, S. Y. S., Motafakkerazad, R., Hossain, M. M. and Rahman, I. M. M. 2012. Water stress in plants: Causes, effects and responses. In: Rahman, I. M. M. and Hasegawa, H. (eds.) *Water Stress.* Rijeka, Croatia: In Tech, pp. 1–14.

Maltby, E. 1991. Wetlands—Their status and role in the biosphere. In: Jackson, M. B., Davies, D. D. and Lambers, H. (eds.) *Plant Life Under Oxygen Deprivation: Ecology, Physiology and Biochemistry.* The Hague, the Netherlands: SPB Academy, pp. 3–21.

Mielke, H. W. and Reagan, P. L. 1988. Soil as an impact pathway of human lead exposure. *Environ. Health Perspect.* 106 (1): 217–229.

Moller, I. M., Jensen, P. E. and Hanson, A. 2007. Oxidative modifications in cellular components in plants. *Annu. Rev. Plant Biol.* 58: 459–481.

Nagajyoti, P. C., Lee, K. D. and Sreekanth, T. V. M. 2010. Heavy metals, occurrence and toxicity for plants: A review. *Environ. Chem. Lett.* 8: 199–216.

Negi, S., Singh, V. Melkania, U. and Rai, J. P. N. 2019. Estimation of heavy metal pollution index for groundwater around Integrated Industrial Estate of Tarai Region. *Environ. Ecol.* 37 (IB): 429–435.

Negi, V. and Singh, V. 2016. A comparative study of lead and mercury in the ambient air in Pantnagar University employing moss *Thuidium cymbifolium* (Dozy & Molk.) Dozy & Molk. *Int. J. Adv. Biol. Res.* 6 (2): 247–252.

Negi, V., Nautiyal, N. and Singh, V. 2018a. *Soil Ecology: The Basis of Sustainable Agriculture and Climate Change Mitigation.* New Delhi, India: Write and Print Publication. 145 p.

Negi, V., Singh, V., Subhpriya, Negi, K., Kumar, V. and Negi, S. 2018b. Heavy metal stress in plants: Cellular and molecular level. In: Singh, V., Negi, V., Kushwaha, G. S. and Negi, S. (eds.) *Environmental Management: Resource and Energy Conservation and Utilization.* New Delhi, India: Best Publishing House, pp. 70–88.

Oono, Y., Seki, M., Nanjo, T., Narusaka, M., Fujita, M., Satoh, R., Satou, M., et al. 2003. Monitoring expression profiles of *Arabidopsis* gene expression during rehydration process after dehydration using ca. 7000 full-length cDNA microarray. *Plant J.* 34: 868–887.

Parmar, S. and Singh, V. 2015a. Phytoremediation approaches for heavy metal pollution: A review. *J. Plant Sci. Res.* 2 (2): 1–7.

Parmar, S. and Singh, V. 2015b. Water quality parameters of River Yamuna in Delhi after 20 years of the Yamuna Action Plan. *Int. J. Sci. Nature* 6 (4): 1–8.

Parmar, S. and Singh, V. 2016. Elemental analysis of chelant induced phytoextraction by *Pteris vittata* using WD-XRF spectrometry. *Int. J. Agric. Environ. Biotechnol.* 9 (1): 107–115.

Raskin, I., Dushenkov, S. and Salt, D. E. 1994. Bioconcentration of heavy metals by plants. *Curr. Opin. Biotechnol.* 5: 285–290.

Rauser, W. E. 1999. Structure and function of metal chelators produced by plants. Cell Biochem. Biophys. 31(1): 19–48. doi:10.1007/BF02738153.

Sachs, M. M. and Ho, D. T. H. 1986. Alteration of gene expression during environmental stress in plants. *Ann. Rev. Plant Physiol. Plant Mol. Bio.* 37: 363–376.

Sachs, M. M., Subbaiah, C. G. and Saab, I. N. 1996. Anaerobic gene expression and flooding tolerance in maize. *J. Exp. Bot.* 47: 1–15.

Sachs, M. M. and Vartapetian, B. B. 2007. Plant anaerobic stress: Metabolic adaptation to oxygen deficiency. *Plant Stress* 1 (2): 123–135.

Satoh, R., Nakashima, K., Seki, M., Sinozaki, K. and Yamaguchi-Shinozaki, K. 2002. ACTCAT, a novel cis-acting element for proline- and hypoosmolarity-responsive expression of the *ProDH gene* encoding proline dehydrogenase in *Arabidopsi*s. *Plant Physiol.* 130: 309–317.

Schützendübel, A. and Polle, A. 2002. Plant responses to abiotic stresses: Heavy metal-induced oxidative stress and protection by mycorrhization. *J. Exp. Bot.* 53: 1351–1365.

Schwartz, D. 1969. Gene fixation studies of anaerobic germination of Adh and Adh+kernels. *Am. Nat.* 103: 479–481.

Seigneuric, R., Mjahed, H., Gobbo, J., Jolly, A.-L., Berthenet, K., Shirley, S. and Garrido, C. 2011. Heat sock proteins as danger signals for cancer detection. *Front. Oncol.* 1: 37.

Spiegel, H. 2002. Trace element accumulation in selected bioindicators exposed to emissions along the industrial facilities of Danube Lowland. *Turkish J. Chem.* 26 (6): 815–823.

Taiz, L. and Zeiger, E. 2002. *Plant Physiology*, 3rd edn. Sunderland, MA: Sinauer Associates. 690 p.

Tatar, O. and Gevrek, M. N. 2008. Influence of water stress on proline accumulation, lipid peroxidation and water content of wheat. *Asian J. Plant Sci.* 7 (4): 409–412.

Vendruscolo, E. C. G., Schuster, I., Pileggi, M., Scapim, C. A., Molinari, H. B. C., Marur, C. J. and Vieira, L. G. E. 2007. Stress-induced synthesis of proline confers tolerance to water deficit in transgenic wheat. *J. Plant Physiol.* 164: 1367–1376.

Yadav, S. K. 2010. Cold stress tolerance mechanisms in plants: A review. *Agron. Sustain. Dev.* 30: 515–527.

Yang, Z. and Chu, C. 2011. Towards understanding plant response to heavy metal stress. doi:10.5772/24204.

Zlatev and Stoyanov, Z. 2005. Effects of water stress on leaf water relations of young bean plants. *J. Central Eur. Agric.* 6: 5–15.

WEBSITES

http://academic.oup.com/jxb/article/52/359/
http://academic.oup.com/jxb/article/54/3588/
https://doi.org/10.1093/jexbot/53.366.1
https://doi.org/10.1093/jexbot/53.372.1351
https://doi.org/10.3389/fonc.2011.00037
https://link.springer.com/article/10.1051/agro/2009050
https://www.agriculturejournals.cz/publicFiles/00477.pdf
https://www.nature.com/articles/srep32965
https://www.researchgate.net/figure/Vacuolar-sequestration-of-heavy-metals-in-plant-cell-Following-the-uptake-through_fig1_221916790
https://www.slideshare.net/dathancs/cold-stress-in-plants

9 Physiological Effects of Climate Change

Let us work together to mitigate the menace of climate change. This would be a great tribute to our beloved Mother Earth

Narendra Modi
Prime Minister of India

LIVING IN AN AGE OF CLIMATE CHANGE

The Globalization Era is gone. The Information Age is gone. We have now ushered in an Age of Climate Threat. Not just an era, but an Age—of Climate Change!

When global warming and climate change had hardly struck our imaginations and touched the human psyche, Henryk Skolimowski (1931–2018) proposed a beautiful idea of the Ecological Age (Skolimowski 1991), which has been elaborated and emphasized by Singh (2019) in his theory of astro-biological evolution: *Fertilizing the Universe*. Global warming and climate-change issues entered into worldwide debates by the end of ninth decade of the twentieth century. By the second decade of the twenty-first century, climate change came out of hot debates, proving itself to be a reality. Yes, climate change is the grimmest reality of our times. A reality that is enlarging itself dynamically, and gradually readying itself to devour the Living Planet. The dire fact facing the contemporary times is that our Living Planet is now stalked in an Age of Climate Change.

The climate generated on Earth with mixed and interacting terrestrial and non-terrestrial factors is a phenomenon that governs the whole life. The earth is a Living Planet, as its climate system is conducive to life. The key phenomenon of life, which is influenced, regulated, maintained, and sustained by climate, is the physiology. A climate system emanating from the thermodynamics of the biosphere holds the temperatures in a specific range in which millions of species and their genotypes—plants, animals, and microorganisms—come into existence, reproduce, develop their natural behavioral characteristics, and sustain vibrantly.

Not all species of all the five kingdoms of life—Monera, Protista, fungi, plants, and animals—exist in all types of climates. In fact, the planet's climate is the average of the sum total of all climate types and microclimates. This "diversity" of climate gives rise to the assemblage of diverse species (with their genotypes) within structural, functional, and self-sustaining systems—the ecosystems. And, thus, a whole diversity of all the earth's ecosystems makes up an integrated, vibrant, and sustainable biosphere—the so-far-known lone biosphere in the universe. Climate variability arises in accordance with geological, geographical, edaphic, atmospheric, hydrospheric, altitudinal, and temporal factors and their interactions, and interactions of all these factors with incoming radiation inputs. Thus, a variety of climatic zones is prevailing all over the earth with each zone spelling its own abiotic and biotic specificities. Each climatic zone again accommodates a variety of ecological zones, and each ecological zone is characterized with its own living inter- and intraspecies biodiversity, ecological processes, and microclimate.

Living species are found in all the climatic zones with varying numbers, genotypes, populations, and densities. It spectacularly occurs despite extreme climatic conditions. Many organisms are adapted to thrive in extremely hot, cold, and humid conditions. There are yet numerous organisms—especially animals and birds—that migrate from harsh to hospitable climates with seasonal changes. An appropriate climate nurtured by thermodynamic regulations, in essence, is a precondition for the life to brood over with all its potential attributes.

An appropriate, soothing, and life-enhancing climate is conducive to human progress. The current trend in climate change is posing a threat to materialistic progress. Agriculture, animal husbandry, industry, transport, navigation, etc. are all set to be disturbed by the changing climate—a response to global warming. This implies that an appropriate climate system is an imperative for sustaining a respectable socioeconomic status of a society, of a nation, and of the whole world.

THE FACTORS AFFECTING THE EARTH'S CLIMATE

Climate change is an outcome of the interactive actions of so many factors, indeed. What is common that is affected and that affects the whole on Earth is the amount of solar radiation received that, in turn, determines the climate system of the planet. There are counts some ten factors, cosmic as well as terrestrial, that are responsible for affecting Earth's climate system (Figure 9.1).

The world has experienced climate change even in the distant past. A cycle of ice ages on Earth has been on record, for example. The changes in the climate in the past, however, have not been as rapid, unpredictable, and baffling as in our times. The root cause of the continuously increasing

FIGURE 9.1   Factors influencing the earth's climate.

processes of environmental disruptions and resulting climate change at an unusual rapid pace is our own species, the *Homo sapiens*. There could be other factors responsible for the ongoing climate change but anthropogenic ones are dominant. The human species throughout the course of its prevalence on the planet has been inclined to transform all natural ecosystems. So stunning have been the dimensions of human actions that today almost all ecosystems on Earth are anthropogenic in nature! Human actions are so extensive and so intensive that a virgin ecosystem is hardly left to be encountered on Earth! Human interventions into natural ecosystems inevitably lead to environmental disruptions. The rate of nature's exploitation by the human species is so vast, so extensive and so abundant that no component of the environment is left undisturbed, intact, and integrated. And environmental disruptions by humans and ominous climate change go hand in hand.

Among the several environmental disruption processes, the one of key significance in the context of climate change is the gradual accumulation in the atmosphere of the gases known as the greenhouse gases (GHGs). The ongoing climate change, in essence, is led by global warming driven by GHGs, dominantly by carbon dioxide (CO_2). Of course, the greenhouse effect on Earth is a natural phenomenon of the biosphere. Without it, the average temperature near the surface of the Earth would be as low as $-18°C$, which would make the planet quite inhospitable for most of the species with the remotest possibility of the human race to evolve and sustain on Earth. It is the greenhouse-effect phenomenon that makes planet Earth what it is: an inhabitable planet. It is the CO_2 gas in the atmosphere that has a central role to play to trap the heat generated by solar radiation and make the environment quite hospitable for all the species prevailing on Earth.

John Tyndall (1820–1893), a physicist, was the first to have ascertained through experiments in 1859 the greenhouse effect imparted by CO_2. Then toward the end of the nineteenth century, a Swedish chemist, Svante Arrhenius (1859–1927), was the first to have calculated the greenhouse effect of CO_2 in the atmosphere, linking it to the ice ages of the geological pasts.

Despite being in traces in the atmosphere, the role of CO_2 in driving climate system is phenomenal, and through photosynthesis and chemosynthesis, and nature's food chains, CO_2 also becomes part and parcel of the whole life of the biosphere. However, accelerated release of CO_2 bound in the fossil fuels (coal, petroleum, natural gas) and living systems (green plants and soils) back into the atmosphere, and uncontrolled production of other GHGs—all due to anthropogenic activities—are the major factors we are concerned about in the context of global warming and climate change in our contemporary times.

The focus on climate-related issues, therefore, is on the processes leading to the emission of GHGs, which have been inarguably exacerbated by human actions. And central to the global-warming process is the atmospheric CO_2 as the global temperature tracks very closely to atmospheric CO_2 levels.

ENHANCED GREENHOUSE EFFECT

The atmospheric concentration of CO_2, the main driver of climate change, reached 410.79 ppm in April 2017, as per the measurements recorded from Mauna Loa Observatory in Hawaii. This atmospheric CO_2 level was the highest in the past 800,000 years for which we have a good record. And yet on May 11, 2019, the new peak of atmospheric CO_2 level recorded at the same laboratory was 415.26 ppm, the highest in the period of the recorded history. And there seems to be no end to the story of CO_2 rise. In the years to come, we are likely to witness new highs of CO_2 in the air, dramatic changes in the composition of Earth's atmosphere, and big leaps of the climate system toward catastrophe.

This CO_2 concentration is no ordinary one; it is an unprecedented advent planet Earth is in the grips of. This (mis) happening might have catastrophic effects on human health and on the planet itself. Average global CO_2 levels over a period of 800,000 years fluctuated between 170 and

280 ppm (Loria 2018), but ever since humankind began burning fossil fuels—since the beginning of the Industrial Age in 1750—things have changed dramatically. It took as many as 1,000 years in a period spanning 80,000 for the CO_2 to mark a rise of just 35 ppm. But in just little over 250 years, we have crossed a whopping 415 ppm mark. And there is no let-up in this increasing trend, which is now 2 ppm per year, which means that in 2100 we shall be breathing in the atmosphere loaded with a CO_2 concentration of 574 ppm. This will cause the temperature to rise to over 6°C.

At what juncture in the prehistoric period has the planet recorded exceptionally high levels of CO_2? This is often debated by climate scientists, and they argue that this might have taken place in the Pliocene Era, about 2–4.6 million years ago, or in the Miocene, about 10–14 million years ago, that is, when the sea-level rise was about 18–24 and 30 m higher than that of today (Loria 2018). However, it is certain that never ever in the past, in any age, has the rise in CO_2 atmospheric concentrations been as rapid as it is being witnessed since the beginning of the Third Millennium. It is why such horrible changes in the thermal behavior of the Living Planet are creating a surge of unprecedented fear among human beings.

At this juncture of the "Climate Change Age" when the average global temperature has risen by 0.8°C since 1880 (according to GISS, NASA's Goddard Institute for Space Studies) with about two-thirds having occurred since 1975 at a rate of about 0.15°C–0.20°C per decade, the world is witnessing rapid melting of glaciers, sea-level rise, increase in superstorms, unpredictable weather patterns, and a spurt in tick-borne diseases. What will be the consequences when there is 6°C rise in the global temperature along with the global CO_2 concentration rising up to 550 ppm? It can be easily predicted: the scenario will be extremely formidable.

The Paris Agreement set the goal of temperature rise up to 2°C or less. The CO_2 levels are climbing continuously and unabatedly, and it appears that the temperatures will soar in the future, and perhaps at an unpredictable pace.

It is not CO_2 alone to blame for the entire dismal climate scenario. The other GHGs, namely methane (CH_4), nitrous oxide (N_2O), chlorofluorocarbons (CFCs), ozone (O_3), and water vapor (H_2O), are adding fuel to the fire. With a gradual increase in the atmospheric concentration of GHGs, the enhanced greenhouse effect is being recorded. In fact, the planet now is stalked in the midst of an enhanced greenhouse effect.

The incoming solar radiation delivers approximately 343 Wm^{-2} of energy at the top of the planet's atmosphere. Out of this amount of energy, about 240 Wm^{-2} penetrates the atmosphere to drive the climate system. Out of the earth's reradiated energy, about 150 Wm^{-2} is trapped by the naturally occurring GHGs: CO_2, H_2O, CH_4, N_2O, and tropospheric O_3. It is the natural greenhouse effect that is vitally indispensable for life on Earth. The post-Industrial Age atmosphere, however, is not characterized just by the naturally occurring GHGs. Anthropogenically generated GHGs over and above the naturally occurring ones, in addition to the special Industrial Age produce, viz. CFCs, add to the natural greenhouse effect, which occurs by means of a direct radiative forcing or climate forcing, that is, a forcing occurring due to the difference between the insolation the planet absorbs and the energy radiated back to space. Influencing the earth's radiative equilibrium, the climate forcing is in addition to what occurs in the earth's natural climate system.

Though a bit controversial, the cumulative effect of the human-added GHGs is about 2.8 Wm^{-2}, or a 1.9% addition to the natural greenhouse effect of 150 Wm^{-2} (Mackenzie 2003). This additional energy attributable to the additional GHGs due to human activities that matters is referred to as enhanced greenhouse effect.

The carbon in the air that is playing havoc with the earth's climate system is also the element that is critical for the generation of the processes leading to life and also an integrated and indistinguishable part of biomolecules in the organisms. Everything organic originates with the fixation of carbon into life, through photosynthesis. "Organic," in fact, is an adjective of carbon. The evolution of life on planet Earth has been possible largely due to the presence of CO_2 in its atmosphere. CO_2, in fact, is a "seed" of life. All other essential factors are to maintain CO_2 "viability." What is life (or part of life) is organic. The role of the inorganic is only to sustain the organic—via photosynthesis and chemosynthesis. Coming from space (from the sun and the other stars in the galaxies) is the radiative energy that synthesizes the myriads of elements and compounds of the environment into life, holds the life in intactness, integrity, and equilibrium. And temperature, a measure of heat energy, in its appropriate range determining weather and climate, ensures the functioning and performance of life.

Temperature, after all, is the crux of all organisms' well-being in the biosphere. With all existing factors that life on Earth reverberates because of, it is the temperature that is phenomenal in brooding an enormous biodiversity at the ecosystem, species, and genetic levels and in striking an ecological balance. In recent decades, however, temperatures are soaring, and the Living Planet appears to be in fever. If the temperatures continue to record this rising trend, the Living Planet might not withstand being in fever for a long time.

OVERRIDING THE EARTH'S POTENCIES

There could be many reasons for the Earth's atmosphere becoming hot as we mentioned earlier. An accumulation of GHGs (all measured against the global-warming potential of CO_2) is the major cause of global warming and consequent climate change and subsequent climate threat. The CO_2 concentration has been regarded as the major indicator of a global-warming-led series of threats. Scientists working with the International Geosphere-Biosphere Programme and the Intergovernmental Panel on Climate Change (IPCC) have demonstrated that humans are the main driver of global environmental changes, eventually leading to climate change. Planet Earth is potent enough to cope with and recuperate the losses its ecosystems suffer by means of self-regulatory processes. Indiscriminate and ever-intensifying anthropogenic activities, however, are gradually overpowering the planet's inherent potencies of ecological regeneration. In other words, it is humanity's load on Earth's ecosystems that determines quality (and deterioration in the quality) of our environment.

Earth's potencies reverberating with self-regulatory processes lie in its lithosphere, hydrosphere, and atmosphere that interact with each other, maintain a state of ecological integrity, and strike a balance/equilibrium necessary for a healthy, vibrant, resilient, and sustainable biosphere (Singh 2019). Human pressure is now overriding Earth's capacity to maintain an ecological equilibrium, and there is no shadow of doubt about it. A major chunk of the human load on Earth's natural resources emanates from the obsession for economic growth, which has become the most robust indicator of a society's or a nation's materialistic progress.

Human civilizations have been evolving within a dynamically stable, healthy, vibrant, and life-enhancing environment. The dynamic environmental stability has been the basis of designing and planning socioeconomic activities, such as agriculture. We have been taming the environment for fulfilling all the needs of our life. And, of course, predictability of environmental stability is the very essence of our happiness and a lure of ushering in a sustainable future (Singh 2019). Our own existence is owing to the environment evolved on Earth over millennia. And, in fact, we make our environment work for us, for our well-being, for our sustenance, for our progress, and for fulfilling all our dreams. We care for our environment (if we do at all!) to draw a benevolent climate out of it.

In our times, thanks to intensifying anthropogenic activities, gradual environmental disruptions are slowly but steadily growing into climate instability. Climate instability, indeed, is a manifestation of environmental disruption. There are some nine dimensions of environmental disruptions, viz. land-use change, biodiversity erosion, nitrogen and phosphorus cycles, global freshwater use, ocean acidification, chemical changes, atmospheric-aerosol loading, stratospheric-O_3 depletion, and climate change (Table 9.1).

TABLE 9.1

Various Dimensions of Environmental Disruptions and Their Consequences

Dimension of Environmental Disruptions	Consequences
Land-use change Conversion of natural forests into croplands, grasslands, parks etc.; Indiscriminate urbanization; Networks of roads, railways.	• Compromise with photosynthetic efficiency of ecosystems; • Adverse effect on biogeochemical cycles; • Reduced resilience and sustainability of ecosystems; • Air, water, and soil pollution; • Erratic water and weather cycles; • Reduces land-filtration capacity and shrinkage in groundwater resources; • Triggering adverse effects of microclimates and global climate patterns; • Frequent occurrence of droughts and floods.
Biodiversity erosion Large number of living species falling into extinction trap	• Reduced photosynthetic/carbon sequestration capacity of ecosystems; • Increased ecological vulnerability; • Decreased ecosystem resilience; • High degree of unsustainability.
Nitrogen and phosphorus cycles Increased flow of nitrogenous fertilizers from industry and spurt in global increase in reactive nitrogen; disturbances in phosphorus cycles	• Erosion of soil biodiversity; • Loss of soil fertility; • Water pollution and eutrophication; • Negative implications for the structure and functioning of ecosystems; • Reduced ecosystem productivity.
Global freshwater use Ever-increasing demand of freshwater for drinking, domestic, agricultural, and industrial purposes	• Reduced water flows in rivers and streams; • Increased stress on aquatic life; • Depletion of groundwater resources; • Adverse impact on global water cycle; • Heavy-metal pollution of groundwater sources.
Ocean acidification Decrease in the pH of the oceans due to increasing uptake of CO_2	• A shift toward decreasing ocean pH; • Inhibition of shell growth and reproductive capacities in ocean animals; • Significant reduction in the ability of reef-building corals to produce their skeletons; • Reduction in the photosynthetic efficiency of ocean ecosystems; • Increase in the possibility of aquatic habitat changes and global hydrological cycle.
Chemical changes As many as 100,000 chemicals in the global market eventually releasing into environment	• Alteration in the physical, chemical, and biological composition of the environment; • Pollution of soil, water, and air; • Poisoning of food chains, adverse impact on ecosystem health, functions, and production flows.
Atmospheric aerosol loading Natural and artificial aerosols adding to environmental pollution load	• Effect on cloud formation and patterns of atmospheric circulation; • Changes in solar radiation reflection and absorption, and thus, on energy budget of the biosphere; • Adverse effect on weather cycles and monsoon system.
Stratospheric ozone depletion Steady decrease in total amount of atmospheric ozone and large springtime decrease in stratospheric ozone over Earth's polar regions due to the catalytic reactions of certain chemicals, especially CFCs and related compounds	• Increase in tropospheric ozone; • Increase in UVB radiation reaching Earth's surface affecting human health, including sunburn, skin cancer, cataract, etc.; • Epidermal damage in animals; • Damage of DNA in plants and animals; • Wiping out of cyanobacteria in the soil; • Retardation of plant growth; • Increase in non-native plant species' invasion; • Damage to biodiversity.

(Continued)

TABLE 9.1 (*Continued*)

Various Dimensions of Environmental Disruptions and Their Consequences

Dimension of Environmental Disruptions	Consequences
Climate change Cumulative effect of all environmental disruptions: A change in global or regional climate patterns attributable largely to increasing CO_2 atmospheric levels	• Negative impact on biosphere's thermal budget: change in weather and water cycles, erratic weather behavior, increased frequency of devastating storms, cyclones, etc.; • Acceleration in snow and ice melt in mountain glaciers and poles; • Unprecedented rise in sea level and inundation of habitats near oceans and seas; • Adverse effects on Earth's ecosystems: changes in the structure, composition, and functioning, alteration in biogeochemical cycles, changes in energy flows, decrease in productivity; • Adverse effects on the physiology and performance of organisms; • Spurt in pathogens and pathogenesis: climate-related diseases; • Acceleration in species' extinction rate.

IMPACT OF CLIMATE CHANGE ON PLANT PHYSIOLOGY

Specialized species and communities prosper within a range of temperatures—that varies over short periods of time (weather) and according to the atmosphere's "behavior" over relatively long periods of time (season). And this becomes the basis for survival, functioning, and sustenance of the species and ecosystems. A range of the temperatures over weather and season—or climate of a geographical region—phenomenally influences life: structure and functioning of communities, which, in turn, are influenced by physiological processes. Deviation in the normal range of temperature, be it in the lower side or in the higher side, induces stress among organisms.

The precipitation regime is gradually becoming a "victim" of the changing temperature regime. The precipitation regime sets up conditions for an evolutionary ecosystem structure, appropriate ecosystem functioning, and a level of primary productivity. Global warming phenomenally activates the hydrological cycle due to which a shift in the precipitation regime can be readily expected, which will create greater "dry versus wet conditions" extremes (Medvigy and Beaulieu 2012). Such a shift is bound to influence soil–water–plant relations and interactions between plants and other organisms. Plant physiological traits, as a result, are readily affected by the changing mode of interactions between plants and other organisms arising out of elevated atmospheric temperatures and subsequent hydrological behavior. Climate change alters species interactions by means of (1) direct effect on mutualists and antagonists, and (2) indirect effect on plant traits that influence the dynamics of these interactions (Becklin et al. 2016).

The probability of species/populations with which they have persisted in the past, are persisting in the present, and would persist in the future amidst changing climatic conditions depends on their three traits, viz., (1) degree of phenotypic plasticity, (2) ability to migrate, and (3) potential to evolve traits adaptive to the changed environmental conditions. Changes in plant physiology and the biotic community influence the strength of selection on plant traits as well as the probability of populations' persistence (Figures 9.2 and 9.3).

Rapid climate change induces the conditions in which a native plant community gets increasingly exposed to altered or "novel" environmental conditions that challenge their physiological limits and adaptation processes (Shaw and Etterson 2012). Earlier, Loarie et al. (2009) suggested that migration of plant populations may not keep pace with accelerated habitat changes at the hands of climate change. Therefore, according to Alberto et al. (2013), rapid evolutionary responses may be

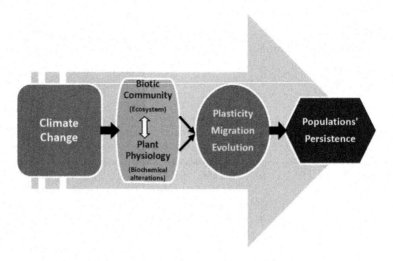

FIGURE 9.2   Effect of climate change on plant physiological traits. (Adapted from Becklin, K. et al., *Plant Physiol.*, 172, 635–649, 2016.)

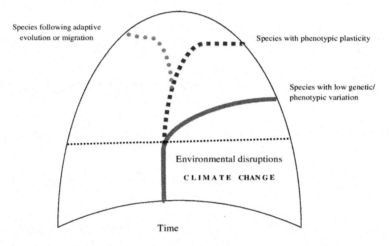

FIGURE 9.3   Probability of species persistence in response to environmental disruptions. (Adapted from Becklin, K. et al., *Plant Physiol.*, 172, 635–649, 2016.)

the dominant processes leading to plants' persistence in the future. Plants tend to evolve a "fitness advantage" in an altered environment by means of attaining physiological plasticity. However, as Anderson et al. (2012) suggest, if the climate changes are rapid enough, the plants may be pushed into "intolerance" ranges despite the most plastic of genotypes.

Rapid climate change will have far-reaching implications even for evolutionary changes both at micro- and macro levels. Microevolutionary responses to these changes may take place over time scales relevant to ecological processes. Macroevolutionary responses involve changes in plant physiology. Lingering adaptations happen to impart limitations on physiological responses to current and future climate changes. Rapid climate change creates a shift in biotic interactions affecting plant traits, fitness to habitat, and even survival. An adverse impact on plant physiology results in altered species with eco-evolutionary consequences (Figure 9.4).

Evolutionary responses to climate change at the physiological level happen to be a key challenge facing plant physiologists and climate experts. A deeper understanding of the potential of the

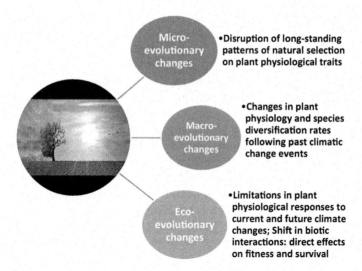

FIGURE 9.4 Micro-, macro- and eco-evolutionary changes due to rapid climate changes.

evolutionary responses would help in predicting plant responses to climate change more precisely. This, in turn, would help the plant physiologists elucidate appropriate measures to respond to the climate-change processes exerting an influence on plant physiology. Dealing with plant physiology requires a combination of individuals and ecosystem processes. The ecological processes are unequivocally sensitive and unambiguously responsive to abiotic environmental factors, and the individuals to ecological processes. For instance, increasing concentrations of atmospheric CO_2 and climate change bring an alteration in photosynthetic rates, which, in turn, shift the plant-growth rate, water-use efficiency, food chains, and ecosystem productivity.

The other examples of plant physiological responses to altered climate patterns emanate from Springer et al. (2008) and Wahal et al. (2013): CO_2 rise increases leaf sugars, which may influence plants' major life-history traits, such as flowering time and fitness induced by sugar-sensing mechanisms.

Physiological pathways have been influenced and determined by the environment the plants inhabit at a certain point of time. Because the environment has not been static over all geological periods in the past, the physiological pathways might not have been precisely the same as the plants exhibit in modern times. The physiological pathways we encounter in the plants today are an outcome of long-term changes in the environment spanning millions of years. Edwards et al. (2010) and Sage et al. (2012) have established this fact basing their studies on photosynthetic pathways. And physiological pathways, in turn, will determine the extent of physiological tolerances for the response to the environments altered by climate-change processes (Becklin et al. 2016).

There is a likelihood that selective environmental pressures on plant physiology exert an influence on plants' abilities to respond to future conditions. Citing an example of the last glacial period, about 20,000 years back when the atmospheric CO_2 level was as low as 180–200 μL L^{-1}, according to Berner (2006), the conditions led to constraining the physiological functioning of C_3 plants. At the contemporary glacial CO_2 level, there is a reduction in photosynthesis and growth in today's C_3 annuals to the extent of 50%, as well as frequent reproductive failure and a high degree of mortality, according to Ward and Kelly (2004).

MICROEVOLUTIONARY RESPONSES OF PLANT PHYSIOLOGY TO CLIMATE CHANGE

Climate change triggers swift alterations in the thermal regime and precipitation regime, together with atmospheric CO_2 levels. As a result, morphology, plant physiology, and life history of a variety of taxa also undergo the process of natural selection. Physiological alterations result in reduced germination success, fecundity, and viability among plants (Anderson 2016). There could be three types of environmental scenarios following rapid climate change: (1) plant species with low genetic and/or phenotypic variability would not persist; (2) plant species with moderate plasticity can facilitate short-term tolerance to climate change; and (3) plant species with a high degree of plasticity can facilitate long-term tolerance/responses to climate change or would migrate to a more favorable environment at higher altitudes/elevations to avoid extinction (Figure 9.3).

Phenotypic Plasticity

The cumulative effects of plasticity throughout a plant's life can be quite extensive (Becklin et al. 2016). For instance, a plant growing under the shade of a canopy would tend to sense a red-to-far-red light ratio below optimum. Such an environmental condition would develop adjustments to enhance light capture via physiological and molecular adjustments (Keuskamp et al. 2010), which, in turn, may induce plastic responses in the subsequent life history of the plants.

Immediate natural-selection pressures that climate change imposes can be alleviated by evolution that is facilitated by plasticity with provision of extended time for evolutionary processes. Adaptive plasticity in water-use efficiency can induce spectacular ability among plants to do well under drought spells and well-irrigated conditions. A striking example is that of the plants from three genetically differentiated populations of *Polygonum persicaria* that, with increased water-use efficiency, fare well under drought as well as irrigated conditions. A befitting response to other environmental stressors becomes possible due to its maintenance under drought conditions. Plasticity can also induce genetic changes in plants (Becklin et al. 2016).

Adaptive Evolution

A population of the plants in an ecosystem facing environmental changes must harbor sufficient genetic variation as well as physiological traits. That is a precondition for the adaptive evolution to take place. Whether a population is replenished or depleted over time would primarily depend on rates of gene flow and mutation (Mitchell-Olds et al. 2007). Lower genetic variation could be detrimental for plant populations. In the event of climate change, the plant populations likely to face high risks of extinction are the ones that are fragmented and contain lower genetic variation. Lower genetic potential of plant populations in a community substantially reduces ecosystem resilience against physical or biological shocks. Species assembled in simple or genetically least-complex ecosystems are likely to be drifted away from the prevailing processes of natural selection. Thus, with the ongoing climate change, the monocultures of fewer species are especially vulnerable to extinction over time.

For the adaptive evolution to be in the folds of genetic variation, there must be a positive relationship between the direction of genetic correlations and that of selection. Additive genetic correlations opposed to the direction of selection are likely to obstruct adaptive evolution against climate change. Climate change can also hamper joint evolution of plant physiological traits. The direction of evolutionary change in the event of climate change would depend on the genetic building up of functional traits in natural plant populations.

Gene Flows

Seed dispersal and pollination are the processes that ensure reproduction, propagation, integration, and sustenance of ecosystems. There are specific spatial connections among plant populations in order to accomplish these natural processes involving gene flows. Gene flows, in fact, are a natural imperative for restricting inbreeding and expanding genetic variations among populations in communities. In addition, gene flows marvelously facilitate evolutionary responses to selection. Genetic enrichment in one population contributes to genetic enrichment in another population. In the natural process of genetic evolution involving substantial gene flows, certain physiological traits—stomatal conductance and drought and frost tolerance, for example—might defy climate change.

Gene flow from genetically rich populations can introduce alleles preadapted to warm conditions in peripheral populations with a narrow genetic base. High rates and unrestricted gene flows may, indeed, be central to evolution of novel environments, fortifying resilience in plant populations and natural communities, creating barriers against ongoing climate change.

MACROEVOLUTIONARY RESPONSES OF PLANT PHYSIOLOGY TO CLIMATE CHANGE

Climate change carries the potential of generating macroevolutionary trends in plant physiology via alterations in patterns of species diversification. Ecological niches in nature would be drastically disturbed, and new niches at their expense would emerge. Vital physiological traits would help the plants provide evolutionary advantage in their new environments. For example, some snow-covered areas in Siberia would turn into areas suitable for agriculture. In the Himalayan Mountain areas, high altitudes would increasingly become suitable for tropical crops. Many a species would shift from their native sites to avail ecological opportunities worth exploiting new niches. In this way, patterns of species distribution as well as of phenotypic evolution would emerge in response to ongoing climate change. The species in their new environments created by climate change may set their geographic range size, fortify population density, and decrease probability of extinction.

The CO_2 concentration–global temperature relationships can be traced back to some 32 million years, from the early Oligocene Age when a significant decline in atmospheric CO_2 concentrations is said to have occurred, which coincided with global cooling and aridification about 14 million years ago in the mid-Miocene Age (Tripati et al. 2009). The global cooling, thanks to a substantial decline in CO_2 levels, led to varying degrees of physiological stress on plants. The plants in the hot and arid conditions were the first to become cold stressed. The physiology of the native plants was substantially affected.

With a decline in the air of CO_2:O_2 ratios and an increase in atmospheric temperatures, there is an increase in the oxygenation reaction with RuBisCO relative to carboxylation (Ehleringer and Monson 1993), resulting in a decrease in photosynthetic efficiency (Becklin et al. 2016). Further, as a consequence, photorespiration scavenges a fraction of the lost carbon from this process. Photorespiration rates and evaporative water loss go hand in hand as higher stomatal conductance is necessitated for making up for carbon losses.

There are physiological pathways that involve CO_2-concentration mechanisms (CCMs), increasing CO_2:O_2 ratios near the CO_2 fixation site, thus resulting in photorespiration decline. C_4 photosynthesis and crassulacean acid metabolism (CAM) photosynthesis are the two striking examples relating to CCMs. Both the CCMs involve separation of initial carbon fixation from the rest of photosynthesis using phosphoenolpyruvate (PEP) carboxylase rather than RuBisCO to fix CO_2 into a C_4 acid. The C_4 acid is subsequently decarboxylated to release CO_2 within photosynthetic cells, where RuBisCO serves to refix it in the Calvin cycle with no, or almost no, photorespiratory carbon losses.

C_4 and CAM pathways make some distinction. In the C_4 pathway, PEP carboxylase and RuBisCO function during the daytime; however, PEP carboxylase is active in mesophyll cells and C_4 acids are transported to bundle-sheath cells where the Calvin cycle goes on. In the CAM pathway, on the other hand, the pattern of stomata opening changes: they open at night and close during the daytime. Thus, PEP carboxylase fixes CO_2 at night, and the C_4 acids are decarboxylated during the daytime, allowing RuBisCO to refix CO_2. Closure of stomata during the daytime makes CAM plants substantially improve their water-use efficiency.

Edwards and Ogburn (2012) suggest that CCMs have evolved numerous times in higher plants over the ages of plant evolution. CCMs, in fact, are the key traits that contributed to increase diversification of certain lineages following the Miocene climate change to the extent that the arid landscapes as of today are dominated by the plants with CCMs.

Two scenarios of phylogenetic patterns of CCM evolution emanating from some studies are as follows:

- Origin of C_4 photosynthesis in grasses, sedges, and eudicots dates to the Oligocene through Miocene (Spriggs et al. 2014).
- Origin of CAM photosynthesis in bromeliads, orchids, and *Euphorbia* spp. dates from the early Miocene to the late Pliocene (Bone et al. 2015).

The greatest advantage derivable from CCMs pertains to the habitats characterized with water paucity, especially with respect to CAM. Going by the contemporary distribution patterns, the evolution of CCMs, according to Sage et al. (2011), took place in semiarid to arid regions. The evolution of C_4 photosynthesis in grasses, according to the study of Edwards and Smith (2010), can be correlated with ecological shifts to open and drier habitats. The evolution of CAM in terrestrial Eulophiinae orchids, according to Bone et al. (2015), can be associated with ecological shifts from humid to hot and dry habitats.

Increased diversification rates in clades as a result of significant shifts evolving CCMs predominantly in the Miocene, as the follow-up of initial CCM evolution, has been intensively studied by Arakaki et al. (2011) (Figure 9.5). Ecological opportunities with species evolving CCMs for diversification, thus, seem to have been created during climate change in Miocene. CCM evolution is associated with "elevated net diversification rates" compared with C_3 plants. Both speciation and extinction rates increase thanks to CCM evolution, suggesting that it is associated with higher rates of species turnover (Becklin et al. 2016).

The CCM evolution may be conditioned by prior physiological adaptations. For example, as Christin et al. (2013) make the claim, in plants of the Poaceae family members, C_4 photosynthesis evolves from the species with an increased proportion of bundle-sheath cells. In the same way, CAM photosynthesis, as Edwards and Ogbum (2012) suggest, may evolve in succulent species because these plants are more capable of storing higher amounts of water and C_4 acids at night.

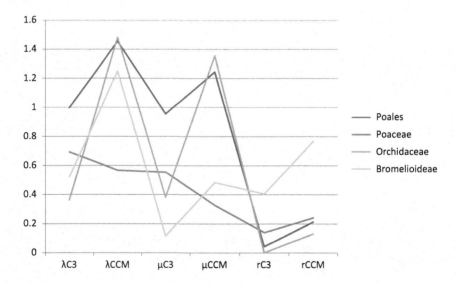

FIGURE 9.5 Estimated speciation rates (λ), extinction rates (μ), and net-diversification rates ($r = \lambda - \mu$) associated with C_3 photosynthesis and CCM-based photosynthesis in clades predominantly in the Miocene, the first geological epoch of warmer climates, some 23.03 to 5.3 million years ago. (Based on the information relating to phylogenetic studies reviewed by Becklin, K. et al., *Plant Physiol.*, 172, 635–649, 2016.)

SPECIES INTERACTIONS AND THE EVOLUTION OF PLANT PHYSIOLOGY IN RESPONSE TO CLIMATE CHANGE

In an ecosystem, plant species evolve amidst the operating environmental factors and their community of interactive populations. In other words, the complex of abiotic and biotic environment of an ecosystem determines how a species is influenced in its path of natural evolution. What is often missed in our discourses on species' evolution in response to climate change is the deviation of interaction among the diverse species in a community. There are three major dimensions of complex interactive behavior of plant species: mutual, competitive, antagonistic. Climate change can bring substantial alterations in mutualism, competition, and antagonisms among plant species in a community and drive the evolution of their physiology (Lau et al. 2014), such as effects of mycorrhizal fungi on nutrient dynamics as well as on carbon balance and the generation of defense chemicals against various pathogens (Becklin et al. 2016). The eco-evolutionary consequences of the indirect effects of climate change relating to alterations in plant interactions might be more effective than even those of the direct effects of climate change (Alexander et al. 2015) (see Table 9.2).

TABLE 9.2

Plant Physiological Traits Evolving Due to Species Interactions in Response to Climate Change

Plant-Herbivore Interactions	Plant-Pollinator Interactions	Mycorrhizal Associations
• Disruption of the production of secondary metabolites leading to lessening of anti-herbivore defense.	• Climate change may affect pollination mutualism through influences on plant physiology.	• Supply of sugars from plants to symbiotic fungi in the rhizosphere may be diminished due to effect of elevated temperatures on photosynthesis and carbohydrate synthesis.
• Decline in the nutritive value of plant tissues.	• Increased CO_2 reduces nutritive value of nectar for pollinators by influencing photosynthesis and sugar production.	
• Increased CO_2 level in air enhances C:N ratio reducing plant leaves' nutritive value for herbivores.	• Elevated CO_2 over nearly two centuries has declined pollen protein concentration, for example, in *Solidago Canadensis* leading to reduced pollen potency.	• Plants' drought tolerance, pathogen resistance, herbivore resistance largely attributable to symbiont fungi are substantially diminished due to disturbances in plant–fungi symbiotic relationship.
• Herbivores need to extract more plant biomass to fulfill their protein demand.		
• Alteration in the selection for plant defensive and tolerance traits.	• Mismatch between flowering stage and pollinator activity due to changes in plant and pollinator physiology.	• Shift in relative resource limitations within host plants leading to alterations in mycorrhizal dynamics and plant role in mutualism.
• Increase in insect population growth rate due to higher temperatures leading to increased intensity and frequency of plant damage.	• Dramatic shift in flowering times and other phenological events due to elevated CO_2 concentrations.	• Disturbance in nitrogen cycle in response to elevated atmospheric CO_2 may strike a competition between host plant and fungi for this scarce resource.
• More frequent and intensive herbivory can alter plant physiology, reduce plant-growth rates, diminish adaptive potential and inhibit genetic diversity.	• Modified patterns in gene flow may affect coevolutionary dynamics between plants and pollinators and decrease seed production.	• Growth depressions in plants due to disruptions in symbiotic relationships.
	• Cascading effects on the evolution of plant traits due to shift in pollination patterns.	• Plants can preferentially allocate carbohydrates to selectively more beneficial mycorrhizal fungi.

Source: Based largely on the literature reviewed by Becklin, K. et al., *Plant Physiol*, 172, 635–649, 2016.

The climate change, in essence, is not occurring or ought to occur, in a linear fashion. There is no exclusive target of elevating CO_2 in the atmosphere and rising temperatures. The climate change has multiple dimensions and triggering effects, and it affects (and further set to affect) every phenomenon of the biosphere including every aspect of living organisms' physiology. The alteration in plant physiology in response to climate change is the root cause of all ecological changes taking place—most likely irreversible. The physiological effects of climate are evolutionary in nature. These ought to dynamically infuse structural and functional changes among communities dependent on photosynthesizers.

EFFECTS OF CLIMATE CHANGE ON FOREST ECOSYSTEMS

Climate change at a global scale is all set to exert its phenomenal impact on the structure, composition, functions, and ecological dynamics of the earth's forest ecosystems. Increasing levels of atmospheric CO_2, global warming, changing precipitation patterns, and climate change being increasingly revealed through escalating frequency and intensity of weather extremes are some of the all-pervading global factors carrying potentially negative implications for the earth's forests. According to a Food and Agriculture Organization (FAO) report (Moore and Allard 2008), these changes in the climate system are already showing notable impacts on the world's forests and the forest sector through: (1) increased frequency of forest fires, (2) a shift of insect species ranges, and (3) longer growing seasons.

There is a natural complex relationship between changing climate, forests, and forest pests that has to determine the state of the forests in the years to come. An understanding of this complex relationship is vital for understanding the ecological dynamics of the forests, the most stable terrestrial ecosystems.

FOREST PRODUCTIVITY

Forest productivity and alpha/species diversity are attributable to many environmental factors, especially temperature, precipitation, and nutrient availability. An increase in the availability of these factors in a forest ecosystem typically increases productivity. Species within the forest, however, might differ in terms of their tolerance to varying environmental factors (Das 2004). The key factor is the temperature that regulates vital ecological processes, viz. litter decomposition, nitrogen mineralization, nitrification, denitrification, nutrient uptake, fine-root dynamics, CH_4 emission, etc. (Norby et al. 2007). A rise in atmospheric temperature, therefore, may change species composition and dynamics of forest ecosystems in several ways. Rising temperature, thus, holds the key to determining forest productivity. The impact of warming will not be uniform throughout a year (Garrett et al. 2006). While in the winter season a moderate warming might relieve plants from cold stress and contribute to an increase in productivity, in the summer season, even a slight increase in temperatures might be detrimental to the forests.

Changes in both precipitation and temperature under the "umbrella" of climate change strongly influence forest moisture regime. A warmer climate brings an increment in evaporation and evapotranspiration rates leading to substantial water losses and a decline in plants' water-use efficiencies. Prolonged warming periods can exacerbate water losses and put the ecosystem vegetation under severe water stress. Such environmental conditions operating in a forest result in reduced primary productivity. However, certain ecosystem characteristics, notably the forest type, age class, heterogeneity, soil depth, and soil organic matter (SOM) would determine the degree of severity of such impacts.

Young plants and saplings are more prone to rising temperatures than the older trees. Being shallow-rooted, the young plants and saplings have to be dependent for moisture from the top soil. Since the top soil is the first to lose water due to increased evaporation in response to a warmer climate, young plants and tree saplings would be the first to become victim of forest desiccation as a result of soil-water loss. Mature trees are deep-rooted and hence can fulfill their water requirements from deeper layers of the lithosphere, thus being less vulnerable to rising temperatures. Forest desiccation is likely to invite more pests and incendiary incidents. Monocultures of even-aged trees and plants inject ecological vulnerability into ecosystems. Climate change further intensifies the damaging effects, such as increased pest onslaughts and fire incidents. A high degree of phytodiversity in a forest, on the other hand, is pivotal in reducing the impacts of climate change. However, even the forest ecosystems encompassing high degrees of biodiversity would fall prey to constant intensification of climate change.

Higher atmospheric CO_2 concentrations could contribute to increase plant-growth rates and water-use efficiency, provided favorable factors, especially water and nutrients, are available in adequate amounts. With a 175-ppm enrichment above the 375 ppm CO_2 level, Norby et al. (2005) have observed multi-year growth increases of 23%. However, positive effects decline with increasing air CO_2 concentrations (Stone et al. 2006). Increased CO_2 enrichment may also come up with changes in plant structure, notably increased number, area, and thickness of the leaf, and a larger diameter of plant stems and branches (Garrett et al. 2006). However, plant responses to elevated CO_2 concentrations would vary from species to species as well as according to local environmental conditions. Increased CO_2 levels, therefore, might bring substantial changes in the composition of a forest ecosystem (Bauer et al. 2006).

Elevated CO_2 concentrations trigger many other environmental changes, such as an increased level of ground-level O_3 (Karnosky et al. 2005), which simultaneously may enhance susceptibility of trees to pathogens (Karnosky et al. 2002), leading to considerable changes in structure, composition, and functions of a forest ecosystem.

Forest Distribution

Distribution of species over a geographical area, altitude, or latitude serves as one of the key indicators of climate change in our contemporary times, and it will be more revealing in the years to come. Warming of the climate is gradually pushing some species toward colder areas/regions. In response to climate change, forest trees and plants are expected to shift northward and/or to higher altitudes, that is, toward the areas with relatively lower temperatures. In the Himalayan region, species occurring at lower altitudes are shifting to higher altitudes and tree line toward the alpine region (Singh et al. 2011). This scenario of vegetation distribution is evidence of formidable climate change.

Lenoir et al. (2008) in their study on 171 forest plant species in Western Europe found that climate warming resulted in an upward shift in species at an average of 28 m per decade. The species that quickly relocate in response to climate warming are the ones with shorter life spans and faster reproduction cycles, for example, herbs, mosses, and ferns. Relocation of larger long-lived tress and perennial herbs, on the other hand, does not take place on a significant time scale. Trees and perennial shrubs in a forest, in fact, are under more severe threat of climate change because they cannot adapt to varying local conditions at a faster pace. Upward movement of alpine flora toward the summit in Switzerland and of tree lines in the Canadian Rocky Mountains, Siberia, and New Zealand have been observed in systematic studies (Parmesan 2006).

Such forest distribution patterns in response to climate change would create forest ecosystem scenarios in a distant future that would be entirely different from those of our present times. The time scale of forest distribution, however, would not be determined by temperatures alone. Other factors, such as edaphic factors, moisture regime, nutrient availability, and human intervention would also exert an impact on the forest species' shifting rates.

Forest Ecosystem Disturbances

All ecosystems are often subjected to various sorts of disturbances. Forests are no exception. Some of the examples are: landslides, invasion of invasive species, strong winds, storms, drought, and pest outbreaks. Such disturbances can change forest structure, composition, and functions. Warming of the climate may influence the timing, duration, frequency, and intensity of such disturbances, which all would exacerbate forest susceptibility to such disturbances. A spurt in forest-fire incidents due to prolonged drought periods can be obvious. An alteration in the disturbance dynamics of forest pests can be readily expected.

There is a complex relationship between climate, forests, and various agents of disturbance. Direct climate change impacts on forests coupled with changes in disturbance dynamics can be substantially damaging to forest ecosystems. A set of disturbances in response to rising temperatures may increase forest susceptibility to other devastating disturbances, such as insect and pathogen outbreaks.

FOREST PESTS

Accelerated insect herbivory is evident from fossil records of prehistoric climate changes due to global warming (Currano et al. 2008). Such evidences from earlier climate-change episodes as well as our current understanding of pest–environment relationships provide enough clues for predicting intensified insect herbivory leading to amplified forest damages in the future.

Climate-change dimensions, such as changes in precipitation and frequent drought conditions, are expected to increase populations, frequency, and damaging effects of forest insects and pathogen pests. Such a change in forest-disturbance patterns would greatly determine the shape and size of the world's forests, as well as of the global-forest sector in near and distant future. Pest populations thriving in a forest ecosystem may respond to climate change in various ways: higher insect fecundity, elevated development rates, range expansion of pests, alteration in host-plant physiology and defense mechanisms, changes in pest–environment relationships, and so on.

Warmer climates are more favorable for the insect pests to proliferate more intensely and feed on plants more profusely than in cold climates. Cold-blooded organisms seek refuge in warmer habitats. Rising atmospheric temperatures will, in fact, be more favorable for insect pests than for larger warm-blooded animals, including mammals. Forest pests can respond to the warmer environment of forest habitats better in terms of higher reproductive rates, high mobility, and enhanced physiology than even small plants. Moore and Allard (2008) refer to the insects and pathogens as the first predictors of climate change.

Temperature is a key factor of climate change influencing the physiology of insect pests (Bale et al. 2002). However, precipitation can also be a crucial factor in the epidemiology of several pathogens, for example, *Mycosphaerella pini* dependent on moisture for dispersal (Moore and Allard 2008). The rise in temperature in the era of climate change will spark other reactions affecting the magnitude of so many other factors, which will induce physiological changes among insect pests. With the physiological response to climate change, the insect pests and plant pathogens may attain greater adaptation capability to rising temperatures. In warmer climates, insect pests may emerge stronger and extract greater amounts of plant biomass. It is more likely that many of the insect-pest species will rule over the "novel" environments created by climate change, and forest plants will be bound to survive under extreme entomological stresses.

IMPACT ON FOREST COMMUNITY

An alteration in the plant–pest–environment relationships is due to the ongoing climate change. This relationship will phenomenally change the mode of competitors and mutualists in a forest ecosystem leading to an alteration in the ecosystem structure and composition of natural communities. Alterations in species abundance and distributions and in phenological patterns, both actual and predicted, are likely to create species alterations which, in turn, are likely to bring modifications in populations' behaviors within communities. Physiological changes within the producers and other organisms within a community are likely to induce drastic changes in the composition and overall functioning of forest ecosystems. With intensification in climate-change processes, a number of pollinators and insect species of high aesthetic values, such as a variety of butterflies, are likely to vanish.

CLIMATE CHANGE AND AGRICULTURE

Humanity's sustainable future, as envisioned today, is tied with agriculture. The world population will increase from about 7.5 billion at present to about 9.7 billion in 2050, according to the United Nations Department of Economic and Social Affairs. At the same time, a decline of about 50% in food-grain production attributable to climate change is likely to take place. An increasing population, followed by decreasing food-grain production, is the most formidable challenge humanity is going to face in the Age of Climate Change. Food security is already at high risks in many areas of the world. According to *The State of Food Insecurity in the World*, a joint report by FAO, International Fund for Agricultural Development (IFAD), and World Food Programme (WFP) (2015), there are as many as 795 million people in the world who do not regularly get enough food to eat. A broad picture of food insecurity, thus, is already visible. A kind of famine seems to be creeping over the globe. An impact of any sort on agriculture may phenomenally affect humanity's designs, plans, programs, dreams, happiness, and future.

Agriculture is both a driver of climate change and a sufferer of climate change. The kind of the food-production system the humanity evolved the world over is not a climate-friendly one. This depends on modified or anthropogenic ecosystems, that is, on cultivated lands. Land till and other agricultural processes generate enormous amounts of carbon emissions. Not only carbon emissions in the form of CO_2 but also emissions of two other notorious GHGs, namely, CH_4 and N_2O, which carry many times more global-warming potential (GWP) than that of CO_2.

Global agriculture sector contributes about 15% to all global carbon emissions, according to the estimates of IPCC (10%–12%), Organization for Economic Cooperation and Development (OECD) (14%), and World Resource Institute (WRI) (14.9%) (IPCC 2007). However, post-harvest processes add fuel to the fire by further adding to emissions. Our food has to undertake a long journey from the cropland to our tables. Before it is consumed, the food grains have to undergo processing, transportation, marketing, fortification, packaging, selling, and cooking. Agricultural operations involving petroleum-dependent technologies and inputs like chemical fertilizers and pesticides and post-harvest operations—from croplands to ending up through consumption—account for about half of the human-generated greenhouse emissions (GRAIN 2011).

The cultivated lands are carved out of forests, rangelands, savannahs, wetlands, grasslands, alpine meadows, and many other virgin lands. The conversion of natural forests into agricultural lands is perhaps the most common land-use scenario of the world, which has been emerging at heavy ecological and environmental costs. According to GRAIN (2011), about 70%–90% of the world's forests are sacrificed only for giving way to agriculture, and as much as 15%–18% of the total global GHG emissions are owing to such deforestation and unwise land-use changes. To keep pace with the ever-increasing world populations, conversion of lush green forests (the major terrestrial carbon sinks) into agricultural lands will continue to be an imminent human activity in future. It means that the terrestrial-carbon sinks will go on shrinking in the future if we do not initiate radical modification in our food-production systems.

Nearly one-quarter of the transportation involves foods and food products. Food processing, transportation, packaging, and retail contribute to about 15%–20% of GHG emissions. In the industrial system of food production, processing, and distribution there is not much concern about food wastes and carbon emissions following biodegradation. According to GRAIN (2011) estimates, 30%–50% of the total foods produced go to waste, adding to 2%–4% of the GHG emissions. Apart from agriculture and food-related emissions, about 43%–56% are non-food-related emissions. Thus, this green revolution-type industrialized agriculture is responsible for 50% of the total human-generated global carbon emissions, ranging from a low 44% to a high of 57% (Figure 9.6).

Current agricultural systems rob the soils of its fertility. Food, in fact, comes from the soil and, ultimately, returns to the soil. Soils of the world, however, are highly abused in the conventional practices of green-revolution agriculture. Cultivated soils of the world, according to some estimates, have lost 30%–75% of their SOM during the twentieth century. These losses from the soils

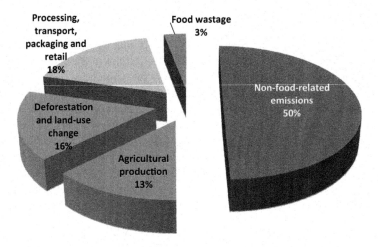

FIGURE 9.6   Agriculture, food, and greenhouse-gas emissions.

of pastures and prairies during this period are of the order of 50%. Global accumulated losses of SOM over the last 100 years are estimated between 150 and 200 billion tons. Not all this SOM ends up as CO_2 in the atmosphere. A significant proportion of the SOM has been washed into rivers and streams and deposited in the bottom of aquatic ecosystems. An estimated 200–300 billion tons of CO_2, however, is released into the atmosphere, that is, some 25%–40% of the atmospheric CO_2 excess is as a result of the mismanagement of the world's soils (GRAIN 2011).

Agriculture, as we have previously discussed, is one of the major causes of climate change. The agricultural systems humanity has designed gradually break down ecological integrity in the following ways:

1. Transformation of complex ecosystems into simpler ones through deforestation (alteration of land use, ecological imbalance);
2. Delinking of croplands from forests/rangelands/grasslands (loss of self-containment feature of agriculture);
3. Erosion of biodiversity by sole dependence on monocultures (fewer plant species and selection of narrow genetic base);
4. Dependence on shallow-rooted, rather than deep-rooted, plants (shift from perennials to annuals);
5. Selection of water-guzzling crop varieties (overexploitation of water resources, soil erosion, soil salinization);
6. Reliance on external (market-based) inputs, for example, chemical fertilizers, mined fertilizers, and pesticides (air, water, and soil pollution, poisoning of food chains, knocking down of soil flora and fauna, and soil-fertility depletion);
7. Dependence on fossil-fuel-consuming heavy machines, for example, tractors and combine harvesters, exacerbating CO_2 emissions (significant contribution to climate change).

All the above-mentioned factors infuse ecological vulnerability in our agriculture. Intensification of this kind of agriculture exacerbates ecological imbalance, leading to further intensification of the processes resulting in the breakdown of ecological integrity. Climate change, in essence, is an outcome of the breaking down of Earth's ecological integrity (Singh 2019).

Farm animals, especially in the developed countries, are largely fed on food grains. For example, as much as 95% of the soybeans produced in the world are fed to livestock, especially to bovine. Producing 1.0 kg of bovine meat releases as much as 200 kg of CO_2 into the atmosphere.

China rears as many as 700 million pigs, that is, one pig per two Chinese citizens. To feed these pigs to produce pork for its people, China imports 80 million tons of soybeans from Brazil. We can now imagine how much CO_2 emission is generated from this single farming activity of pork production (Perrone 2018).

A large number of farming communities in the world, especially in the mountains and highlands, are livestock dependent. Livestock may not be big players in the world's economy, but they are critical for the livelihoods of billions of people. However, at the same time, they happen to be major actors in accelerating climate-change processes. Livestock are also among the worst sufferers of climate change. They are far more vulnerable to climate change than cropping systems. Crossbreeds in the South Asian countries are more vulnerable than their indigenous counterparts (Singh et al. 2017).

Livestock-sector contributions to the total GHG production appears to be extremely formidable, indeed. The livestock–climate change linkages can be understood following Figure 9.7. Considering emissions along the entire commodity chain, livestock's contribution to global warming is about 18%. Some of the major FAO findings are presented in Table 9.3. Of the worldwide emissions, that is, 7,516 million metric tons per year of CO_2 equivalents (CO_2e), or 18%, are contributed by cattle, buffaloes, goats, sheep, camels, horses, pigs, and poultry (Steinfeld et al. 2006). World Watch analysis, however, shows that livestock and their by-products actually account for at least 32,564 million tons of CO_2e per year, or 50% of annual worldwide GHG emissions (Goodland and Anhang 2009) (Table 9.4).

The widely accepted GWP value for CH_4, using a 100-year time frame, is 25. However, as Goodland and Anhang (2009) suggest, the GWP value of CH_4 is 72 using a 20-year time frame, which appears to be more appropriate. Livestock contribute about 9% of total CO_2 emissions, but 37% of CH_4 and 65% of N_2O (Singh et al. 2017).

Livestock waste management at the global scale leads to the generation of about 9.3 Tg/year CH_4 (Scheehle 2002). Contribution of the developed countries to this amount of CH_4 production is as high as 52%. Livestock waste management in India is better than that in the Western countries.

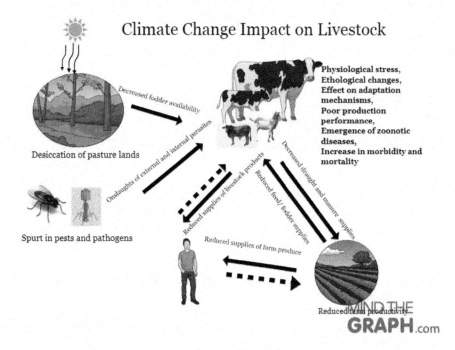

FIGURE 9.7 Livestock–climate change linkages. (From Singh, V. et al., *Indian J. Anim. Sci.*, 87, 11–20, 2017.)

TABLE 9.3
Livestock Contributions to Environmental Perturbations

Land and biodiversity	• The only human-managed user of land;
	• More than one-fourth of the ice-free area of the earth used for grazing;
	• One-third of the total cultivated area used for feed/fodder production;
	• Largely responsible for forest degradation;
	• Degradation of about one-fifth of the pastures due to overgrazing;
	• Breeding and maintenance at the cost of wild animals in terms of the land area they occupy and plant biomass they consume;
	• Responsible for deforestation leading to biodiversity erosion;
	• Facilitators for invasion by alien species;
	• Threat to about 37% of the terrestrial eco-regions;
	• As many as 35 global biodiversity hotspots facing threat of habitat loss.
Water	• About 8% of global water used for irrigation of fodder crops;
	• One of the largest sources of water pollution;
	• Animal wastes as a major source of water pollution leading to eutrophication of natural water sources.
Atmosphere and climate	• Contribution to 18% CO_2-equivalent GHG emissions;
	• Livestock sector accounts for 9% of anthropogenic CO_2 emissions, 37% of CH_4, 65% of N_2O emissions, and 64% NH_3 emissions contributing;
	• Significant contribution to acid rain and acidification of ecosystems.

Source: Steinfeld, H. et al., *Livestock's Long Shadow: Environmental Issues and Options*, FAO, Rome, Italy, 2006; Singh, V. et al., *Indian J. Anim. Sci.*, 87, 11–20, 2017.

TABLE 9.4
Uncounted, Overlooked, and Misallocated Livestock-Related GHG Emissions

	Annual GHG Emissions (CO_2e)	Percentage of Worldwide Total
FAO estimate	7,516	11.8
Uncounted in current GHG inventories:		
1. Overlooked respiration by livestock	8,769	13.7
2. Overlooked land use	≥2,672	≥4.2
3. Undercounted CH_4	5,047	7.9
4. Other four categories	≥5,560	8.7
Subtotal	≥22,048	≥34.5
Misallocated in current GHG inventories:		
5. Three categories	≥3,000	≥4.7
Total GHGs attributable to livestock products	≥32,564	≥51

Source: Compiled from Goodland, R. and Anhang, J., *Livestock and Climate Change: What if the Key Actors in Climate Change were Pigs, Chickens and Cows?* Worldwatch Institute, Washington, DC, 10–19 pp, November–December, 2009

It is because Indian breeds lead to comparatively much lower levels of CH_4 production. With GWP equivalent to 296, N_2O production from manure management in India are projected to be 0.022 Tg/ year by 2020 (Sirohi and Michaelowa 2007).

The effect of agriculture on climate change is reciprocal. That is, climate change imparts its phenomenal impact on agriculture. And there are multiple dimensions of the climate-change effects on agriculture, ranging from the quality to the quantity of the production flows from agricultural systems.

Agriculture and climate are inextricably linked with each other. The environmental factors that are being phenomenally influenced by climate change (e.g., Earth's temperature, heat waves, precipitation, hydrological cycle, and nutrient cycles) are also phenomenally influencing agriculture, including soil health, water use, photosynthesis, crop yields, and agro-biodiversity.

Water is fundamental for ecological phenomena, including photosynthesis. Water availability in appropriate amounts as per the physiological needs of crop plants and for the maintenance of soil health is a primacy for the performance of agricultural systems. Withdrawal of water for longer periods (water stress or drought) and excess of water (floods) exert stress on crop plants. The hydrological cycle is pivotal in recharging natural water resources (rivers, streams, ponds, groundwater) and ensuring continuous water supplies for crop production. Hydrological changes are already resulting in intense and frequent droughts and floods. With further soaring of temperatures, weather cycles will become more erratic and drought-flood cycles will be more damaging to agriculture. Most of the regions of the earth will soon be in the grips of climate change, provided the current trend in temperature rise continues unabated. A significant decline in crop yields, increased frequency of crop failure, food shortage, and famine would then be common scenarios in our world.

Temperature is the major determinant and regulator of plant physiology. Excessive heat (high temperatures), one of the major characteristics of climate change, would jeopardize the thermodynamics of the biosphere. Many of the crops and/or crop varieties in an agroecological region may cease to perform. And yet, some crop cultivars may survive well if shifted to the agroecological zone in a comparatively cooler region. The quality of certain cereal crops would deteriorate. For example, the Basmati rice would not develop its specific aroma if it has no exposure to lower temperature toward its maturing stage.

Independent of other variables due to climate change, the steadily increasing atmospheric CO_2 level will have a profound direct effect on physiology, biochemistry, and growth of food crops. It is projected that CO_2 concentration would continuously rise to as much as 500 to 1,000 ppm by the year 2100 (IPCC 2007) (Figure 9.8). C_3 plants, which account for 95% of all the plants, such as wheat, rice, and soybeans, would respond more conspicuously to elevated CO_2 levels than C_4 plants, such as maize, sorghum, and sugarcane. It is because C_3 plants take up CO_2 directly from the atmosphere and, therefore, elevated CO_2 might be beneficial for the plants. C_4 plants, on the

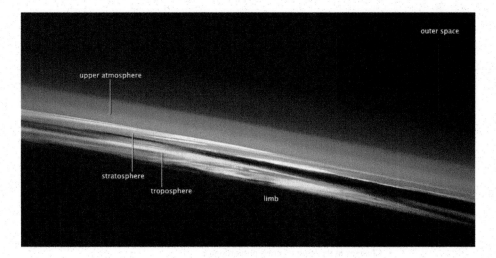

FIGURE 9.8   Sunset from the International Space Station. CO_2 concentrations in the earth's atmosphere (troposphere) are continuously rising. It is projected that CO_2 concentrations would continuously rise to as much as 500 to 1,000 ppm by the year 2100 (IPCC 2007). Photosynthetic assimilation of CO_2 is central to the sustenance of life, and atmospheric CO_2 concentrations to the very fate of the biosphere. (Courtesy of ISS Crew Earth Observations and Image Science & Analysis Laboratory, NASA Johnson Space Centre, Houston, TX.)

other hand, first concentrate CO_2 within and produce malate inside the plants, which then enters the photosynthetic cycle. Thus, C_4 plants respond poorly to elevated CO_2. Although elevated atmospheric CO_2 causes increased photosynthesis in plants (Thomson et al. 2017), a greater production of carbohydrates and biomass will occur in C_3 than in C_4 plants.

There is, however, no linear relationship between gradual CO_2 rise and photosynthetic rates even in C_3 plants. After a certain level of CO_2 increase, the photosynthetic rates level off. It might be because with no corresponding increase, the other necessary inputs of photosynthesis, that is, light and water, may become limiting factors. A greater intensity of light but limitation of water would lead to closure of stomata, cutting off CO_2 supplies to the plants. An adequate supply of water but lower light intensity will also not be realized in enhanced photosynthetic rates despite the availability of greater amounts of CO_2 in the atmosphere.

Unabated climate change may also destroy crop biodiversity. What is of vital importance in crop biodiversity is a variety of wild crop relatives, which serve as key genetic resources for the breeding of crops in the future. Distribution of the wild-crop relatives will be severely affected due to climate change, which would be a jolt to farming in the future. There is a likelihood that all varieties of food crops would not be able to adapt to the changing environment during the climate-change regime. Only a few varieties of fewer crops may adapt to the changing climate while most of them would vanish. Thus, rising temperatures would impose increased reliance on monocultures, which means a compromise with the principles of sustainable agriculture.

Many other indirect climate-change impacts will include undesirable alterations in agroecosystems. Changes in the soil microflora, microfauna, mesofauna, and macrofauna, and in vegetation composition, would invite biotic stresses, like competitions due to proliferation of weeds and suitable conditions for invasion by exotic plants. There can be a spurt in the populations of pests and pathogens (including expansion of ranges and seasons) that may cause large-scale crop damages.

Animal husbandry suffers more from climate change than crop husbandry. Table 9.5 summarizes many dimensions of the farm animals' suffering from climate change.

TABLE 9.5
Some of the Impacts of Climate Change on Farm Animals

Environmental Resource	Impacts

Pastures and feeds/fodders	• **Land use and systems change**
	  • Alterations in species' niches (plant and crop substitution);
	  • Modifications in animal diets;
	  • Decrease in smallholders' abilities to cope with feed deficits.
	• **Changes in the primary productivity of crops, forages and rangeland**
	  • Changes in harvest indices for food–feed crops;
	  • Decrease in the nutritive values of straw;
	  • Decline in the availability of metabolic energy for dry season feeding;
	  • Substantial decrease in rangeland productivity.
	• **Changes in species composition**
	  • Changes in optimal growth ranges for different species;
	  • Alteration in species competition dynamics;
	  • Drastic changes in the composition of natural grasslands;
	  • Increase in the proportion of browse in rangelands due to increased CO_2 levels;
	  • Changes in the legume–grass proportions.
	• **Quality of plant material**
	  • Increase in the lignification of plant tissues due to increased temperatures, resulting in decreased digestibility;
	  • Adversely affect smallholders' incomes and food security due to decline in livestock production;

(Continued)

TABLE 9.5 (*Continued*)
Some of the Impacts of Climate Change on Farm Animals

Environmental Resource	Impacts
Biodiversity	• Loss of genetic and cultural diversity in agriculture, in crops as well as domestic animals (Ehrenfeld 2005); • As many as 20%–30% of all plant and animal species to suffer from major losses with 2.5°C increase in global temperature above pre-Industrial levels (IPCC 2007).
Water	• Increasing water scarcity; • Decreased availability of feeds/fodders as a consequence of water scarcity.
Animal (and human) health	• A spurt in vector-borne diseases, such as malaria, animal tick-borne diseases, bluetongue, etc.; • Increased incidences of large outbreaks of disease (e.g., Rift Valley Fever virus in East Africa); • Increase in helminth infections; • Effect on trypanotolerance in subhumid zones of West Africa; • Increase in the effects on distribution and impact of malaria in many systems and schistosomiasis and lymphatic filariasis in irrigated systems (Patz et al. 2005); • Increases in heat-related mortality and morbidity (Patz et al. 2005); • Increase in susceptibility to HIV/AIDS as well as to other diseases (Williams 2004).

Source: Based on Singh, V. et al., *Indian J. Anim. Sci.*, 87, 11–20, 2017.

COPING WITH CLIMATE CHANGE

Coping with climate change is the most pressing need of our times. Putting climate back in order is not only necessary but an imperative. It is not yet too late to work out all workable strategies and tactics to bring the climate pattern back on track. Since temperature rise is not uniform throughout the globe, the climate-change pattern is also not uniform. If a rise in temperature is socioeconomically beneficial in some cold regions of the world, the benefits are not going to be sustainable. Short-term gains cannot outweigh long-term gains. So, we have to take stock of the long-term socioeconomic scenario to emerge out of the persisting (and intensifying) changes in the climate pattern. We need to be wary and watchful about the climate changes, as we already are, and chalk out three strategies to deal with the emerging scenarios. These strategies involve three actions:

1. Preparedness
2. Adaptation
3. Mitigation

PREPAREDNESS

If climate change begins translating into adverse effects on the socioeconomic state of the people in an area or a region, what is our preparedness to cope up with the situation? Preparedness is a short-term measure to fight against the situation arising out of climate change. Nevertheless, it is very crucial part of the strategy. Every cautious society, a governing body, and a state has its own preparedness against any adversity, against any natural disaster. However, there seems to be no such preparedness against climate change. A few countries of the world have begun declaring a "climate emergency," which itself is an effort toward being conscious before the climate crisis hits hard. Declaring a climate emergency is definitely the first step toward formulating a "preparedness" strategy to combat climate crises. But there is no global consensus on this issue. Even disaster-management strategies are grossly lacking preparedness against climate change.

The climate-strategy focus so far is merely on resorting to long-term measures, viz. adaptation to and mitigation of climate change. Since short-term measures to alleviate or avert or heal up the impact of climate change are necessary, preparedness is also one of the key strategies to deal with the climate change. A climate emergency declared at the global level will greatly help formulate an effective preparedness plan to immediately deal with the climate disaster experienced in an area, country, or a region.

ADAPTATION

We cannot come out of the mess created or to be created by climate change. So, we have an option to be in tune with climate change, that is, to evolve adaptation strategies to avert or minimize adverse effects of climate change on socioeconomic processes. Our conventional agriculture is vulnerable to climate changes. The first direct effect of climate change is on our food supplies and food security. To adapt to the persisting and intensifying climate changes, our strategies should involve:

1. Development of ecologically more stable and productive agroforestry systems
2. Selection of the seeds that need less water and are tolerant to drought conditions and rise in temperatures
3. Genetic development of seeds resistant to climate changes
4. Increasing dependence on landraces specific to agroecological zones that are resistant to adverse climatic conditions
5. Cropping systems comprising deep-rooted and shallow-rooted crops (combination of cereals and leguminous crops, for example)
6. Increased dependence on no-till farming that ensures minimum carbon emissions and soil and water conservation
7. Development of soil and water conservation oriented and efficient irrigation agrotechniques (drip irrigation, for instance)
8. Development of and increased reliance on ecologically sound and environmentally safe organic agriculture

Our modern agriculture is extremely vulnerable to climate change. However, in many remote and less-accessible regions, such as in the Himalayan Mountains, there are many landraces of cereals, millets, and pseudocereals that are well-adapted to harsh and inhospitable environments and adverse climate conditions. In the Himalayas, there are many landraces of rice that thrive well even under unirrigated or rain-fed conditions and are on par with the so-called high-yielding rice varieties in their yields. Millet crops like finger millet (*Eleusine coracana*) and barnyard millet (*Echinocloa frumentacea*) are drought tolerant. Pseudocereals, especially *Amaranthus*, grow well in poor, acidic, degraded soils and require no irrigation. These crops are often better than the "improved" ones in terms of their nutritive values and other traits like pleasant aroma and medicinal values. An increased dependence on such local crops and landraces and their unique genetic material for propagation will be vital for living in tune with climate change.

MITIGATION

Climate-change mitigation is a long-term strategy and involves the formulation of policies, programs, projects, missions, and goals at the local, national, regional, and global levels. There can be many dimensions of the strategy, for example, cutting down, minimizing, or completely averting carbon emissions; development and dependence on clean sources of energy; and accelerating carbon-sequestration processes. Among them, we shall focus on carbon-sequestration-enhancing processes that involve the role of photosynthesizers in natural and anthropogenic ecosystems and socioeconomic systems, especially agriculture. We can call these potential strategies of climate-change mitigation botanical strategies. These are subsequently discussed.

CLIMATE-SMART PLANET: A SUSTAINABLE PLANET

There is no shadow of doubt that climate change is a dire reality of our times. The kind of future we shall usher in with the dismal state of the climate in our present times is well predicted. The future is likely to be grim. Climate change is pushing all hopes and lure of sustainability into an abyss of despair and darkness. As human populations of the world, or say, the human species, is largely, rather almost exclusively, responsible for this menacing climate change, the solutions have also to be anthropogenic. Climate-change mitigation strategies are there in our theories, in our discussions, in our arguments. Should they be implemented on ground, the dismal scenario will certainly transform into a bright one. Different components of the planet's environment must be involved in our strategy, and photosynthesis must be the central phenomenon of our tactics of building up a climate-smart planet.

ECOLOGICAL INTEGRITY AND ECOLOGICAL SECURITY

Ecological integrity emanates from a universal ecological law: everything is related with everything else. Nothing in the living systems can be isolated from the rest. All is one, as Henryk Skolimowski's philosophy says (Singh 2019). Ecological integrity, in fact, is the ultimate truth pertaining to our Living Planet. It is not a static state of life. It is a dynamic phenomenon ensured by continuous flows of matter (nutrients, water, and gases) and energy. Biogeochemical cycles are vital for maintaining ecological integrity of the biosphere. Energy in the form of solar radiation (with a fraction of radiation from other stars of the Milky Way) enters into the biosphere and holds the whole life into integrity. Millions of species, the living forms, exist in inter-relationships, which are vital for ecological processes.

There are two bases of the ecological integrity: physical and physiological. The physical basis includes light, temperature, water/moisture, minerals, atmosphere, and meteorological factors, which are all intertwined into a phenomenon—the climate. The physiological basis involves photosynthesis, the core phenomenon of the biosphere. All structural, morphological, biochemical, and functional attributes are intertwined into a phenomenon—the life. The physical basis and the physiological basis together contribute to evolve and sustain ecosystems. All ecosystems integrate into a biosphere. And the biosphere flowers with a variety of life. In the biosphere, matter flows (biogeochemical cycles and nutrient recycling), energy flows (in trophic levels, from producers to top carnivores), and an ecological equilibrium is maintained. The biosphere holds itself into a state of dynamism commended by ecological integrity.

Ecological integrity can be understood at the level of an ecosystem—a forest ecosystem, for example. The soil of a natural undisturbed climax forest is a wonderful invisible world of microorganisms. There is, in fact, a world far more fascinating than the visible above-soil world. The nitrogen-fixing bacteria in the soil habitat fix nitrogen—one of the structural and functional components of living organisms—into life. All life-constituting mineral nutrients and water flow into the plant's body through plant roots imbibing in the soil. The atmosphere has the two vital gases (CO_2 and oxygen) that flow through the leaves of the plant. Energy (sunlight) coming from the Sun is trapped by the green leaves containing chlorophyll. Energy, nutrients, and gases flow throughout the body of a plant and are functionally bound together by the phenomenon of photosynthesis in which carbon is fixed into sugars, and light into energy of life (the biochemical energy). Through physiological processes, more and more complex biomolecules are synthesized with genetic control over the physiology of the organisms.

All these nutrients, water/moisture, and gaseous flows occur continuously within the ecosystem and, in this way, throughout the biosphere. What will happen if these flows within the biosphere are obstructed? The very phenomenon of life, photosynthesis, will be diminished. Physiological processes among organisms will become impaired. The richness of life will decline. The very dynamism of the biosphere will be disturbed, and ecological balance will be at stake. Ecological integrity of the biosphere will begin breaking down. The ongoing climate change, in fact, is an outcome of this continuous breakdown of the biosphere's ecological integrity. The physical aspect of the ecological integrity that manifests into climate affects the overall physiology of the biosphere. As a consequence, structure, composition, and functioning of the biosphere are severely affected.

There is a reciprocal relationship between the physical (climatic) and physiological aspects of the biosphere's ecological integrity. The physical aspects and the physiological aspects influence each other and are influenced by each other. The balance (or imbalance) between the two determines the state of the ecological integrity. There is one element of this relationship in human hands: control of the photosynthesizers in terms of space provision and resource extraction rates. If the plants are allowed to flourish as forest ecosystems on larger geographical areas, and if aquatic ecosystems are managed properly not allowing them to become polluted, ecological processes will be boosted phenomenally and in due course of time will culminate into an ecological balance and then into ecological integrity.

LIVING SOIL TO NURTURE THE ROOTS OF LIFE

Let you reach for the stars,
But your roots should be deep in soil.
If you own soil,
You will have command on the skies.

(Singh 2019)

Roots of all human civilizations lie in the thin layer of soil that covers the entire land surface of the earth. The health of all the terrestrial beings—microorganisms, plants, animals, human beings—cannot be separated from the health of the world's soils. A few centimeters thick, the layer of the soil accumulated on Earth's surface over the eons, provides a healthy medium for the plants to grow and prosper. Plants, in turn, provide a protective cover to the soil and fortify soil health. The soil–plant relationship is reciprocal, as the health of the one is ameliorated by that of the other. The mutually supported ecosystem health further ameliorates the health of all the terrestrial organisms.

In the last century, especially since the inception of the so-called Green Revolution agriculture, soil-erosion rates have begun exceeding the soil-formation rates. Primitive agriculture largely relying on forest-based or uncultivated foods (wild fruits, pods, buds, flowers, stems, leaves, underground stems, roots, mushrooms, etc.) and traditional agriculture with strong forest–cropland linkages depended on soil health as their most precious natural capital. In those agricultural systems, farmers upheld a reverential attitude toward soils. Soil was regarded as a living system. A farmers' adage says: feed not the plants; feed the soil, so that soil itself feeds the plants. The Green Revolution agriculture derecognized the soil-as-an-ecosystem concept. Chemical fertilizers became a mantra for taking care of soil fertility, and high-yielding varieties (HYVs) of food crops were regarded as a key to agricultural productivity. In order to acquire and sustain the potential yields of the HYVs, more and more chemical fertilizers and heavy irrigation became inevitable. Intensive land tilling using heavy tractors and excessive water use in HYV crops accelerated soil erosion, and HYV seeds robbed the soils of their essential nutrients, wiping out soil microflora and fauna. SOM and soil organic carbon (SOC) were hardly of concern in the Green Revolution. Continuous negligence of soil-fertility management and large-scale deforestation, mostly for cultivation purposes, led to accelerated soil-erosion rates.

Soil is one of the largest and the most wonderful ecosystems on Earth. Soil is the foundation of civilizations. Soil is the key to a sustainable future. Soil life is far more fascinating than the life above ground. Most of the protists, monerans, and fungi in the biosphere inhabit soil. Soils provide a refuge to a diversity of life including bacteria of all shapes and sizes (minute spheres, spiral threads, cylindrical, filamentous chains, and flagellated rods). In fact, there are far greater varieties of bacteria in a variety of soils than any other organisms on Earth. Numerous animals of varying sizes, like ranging from the larvae from numerous insects and earthworms to large ones like reptiles and mammals, such as rodents, find the soils as their most appropriate habitat. Soil, thus, is a home to microorganisms, mesoorganisms, and macroorganisms. A teaspoon of native grassland soil contains between 600 and 800 million individual bacteria that are members of, perhaps, 10,000 species. Several miles of fungi are in that teaspoon of soil, as well as 10,000 individual protozoa. There are 20–30 beneficial nematodes from as many as 100 species (Hargesheimer 2007; Singh 2019).

Communities of microorganisms also vary according to the depth of the soils. Top soil is the most vibrant and active layer harboring as many as 7,800,000 aerobic bacteria, 1,950,000 anaerobic bacteria, 2,080,000 actinomycetes, and 119,000 fungi per gram of soil up to just an 8-cm depth. The microbial populations go on decreasing significantly with an increase in soil depth to the extent that at a 135-cm depth, the aerobic bacteria count is just 100, anaerobic bacteria 400, and fungi 3,000 per gram of soil. Actinomycetes are almost absent at this depth. Such is the invisible but astounding life blossoming within the soils of the world!

Carbon is the basis of life on Earth. All living organisms depend on the supply of necessary elements from the earth. The terrestrial carbon cycle is largely dominated by the balance between photosynthesis and respiration. Carbon is transferred from the atmosphere to the soil via "carbon-fixing" autotrophic organisms, mainly photosynthetic plants and microbes (Lu and Conrad 2005; Trumbore 2006) that synthesize atmospheric CO_2 into organic material. Microbes are critical in the process of breaking down and transforming dead organic material. The organic matter is composed of sugars, proteins, amino acids, nucleic acids, lignin, waxes, celluloses, etc. The degradation of these complex compounds into ionic form is carried out by soil microorganisms either directly or indirectly by the process called mineralization that can be reused by other plants and soil biota.

SOM holds more than 95% of soil nitrogen, 5%–60% of total phosphorus, and about 30% of soil sulfur. The availability of these nutrients is conditional to decomposition of organic matter by microorganisms. Thus, major nutrient cycles (C, N, P, S) are interconnected to maintain homeostasis between the above- and below-ground network.

What is productivity? Productivity is not something that soil automatically expresses into, and it is not an "end product" of the soil. Productivity is the biophysical transformation of ecosystem functions. Productivity is an attribute of ecosystem function, in essence (Singh 2019). Photosynthesizers in an ecosystem capture solar energy and transfer it into biochemical (food) energy. Efficiency of an ecosystem to capture solar energy and transform it into food energy depends upon the health of its soil. The greater the richness of photosynthesizers and biodiversity in an ecosystem, the higher the capacity of the terrestrial ecosystem to transform solar energy into food energy.

The pedosphere, the sphere of the soil, is biologically the richest and the most active portion of the biosphere. The larger the populations and the greater the diversity of the microorganisms and mesoorganisms in the soil, the higher the degrees of soil fertility and, consequently, the higher the productivity of the ecosystem. Soil biodiversity, fertility, productivity, and sustainability are directly related and have bearing on each other.

Soil biodiversity is largely attributable to carbon sequestration through photosynthesis and is vital for maintaining an appropriate climate. By supporting a rich biodiversity of photosynthesizers within and above ground, a healthy soil is directly linked with enhanced photosynthetic efficiency. Soil health imparts a healthy environment for the above-ground biodiversity, contributing to the reduction of global warming and the subsequent mitigation of climate change. Serving as one of the largest natural carbon sinks, the soils of the globe store more than twice as much carbon as the plant biomass of the whole world (Singh et al. 2014). One ton of soil carbon is equivalent to 367-ton atmospheric CO_2. Soil is, thus, a wonderful and fascinating ecosystem on Earth! Roots of the climate-smart and sustainable planet lie in healthy soil. Douglas Jerrold rightly says: he, who owns the soil, owns up to the sky.

Forests for a Healthy, Vibrant, and Sustainable Planet

Forests on the land surface of the earth are one of the most significant components of the biosphere for cooling the earth. A forest is not just a random assemblage of trees in a certain geographical area. It is a biotic community representing a unique biodiversity of plants, animals, and microorganisms with trees being the dominant members of the biocommunity. A natural forest is an ecosystem defining ideal life-infusing and life-enhancing relationships between abiotic and biotic factors phenomenally influencing the thermal dynamics of the biosphere. In addition, a forest plays a crucial role in the moisture circulation/hydrological cycle, nutrient cycles, carbon sequestration, gaseous flows, food chains and energy flows, pedogenesis (soil formation), soil and water conservation, protection against harmful radiation and atmospheric pollution, biodiversity protection, conservation and enhancement, weather cycle, and climate regulation. Natural climax forests also serve as "cradles" of new species' evolution.

There were times when most of the earth's land surface was covered by natural forests. Now the planet's forest cover has recorded its shrinkage to a formidable extent. Most of the world's forests are no more in a state of ecological climax. Human intervention almost in the entire world's forests has hardly left an area under virgin forests. Degraded forests, fragmented forests, an assemblage of thin strands of even-aged trees or monocultures, etc. which we are increasingly witnessing in our times, are not forests as per their definition. Even such modified forests play a role conducive to environmental conservation, but significantly less crucial than the natural and ecological climax forests.

Every tree serves to do what an assemblage of trees or a natural forest does. But its contribution is too small to make a significant difference in the quality of the environment. There are some tree species—*Ficus religiosa* and *Ficus benghalensis* (Figure 9.9), for example—that in their individual mature stage serve as an ecosystem. A glaring and baffling example is that of the Great Banyan Tree located in Acharya Jagdish Chandra Bose Indian Botanic Garden, which is ecstatically spread over a 1.5-hectare area. A single tree akin to a forest ecosystem in itself! A forest community functions in a coherent and integrated way, and its contribution to environmental quality

FIGURE 9.9   *Ficus benghalensis.* (Courtesy https://www.freeimages.com/photo/banyan-tree-1516030.)

maintenance is phenomenal. However, with trillions of trees on the lands of the planet, in addition to the present forests (natural as well as degraded), will play a phenomenal role in enhancing the carbon-sequestration phenomenon for the climate-change mitigation.

A tree is a living representative of solar power on Earth. A tree can bring the Sun down to Earth. Of course, all plants containing chlorophyll are empowered to bring the Sun down to Earth, but trees are part of a system enacting this phenomenon more efficiently, more vibrantly, and more sustainably. Bringing the Sun down to Earth literally means photosynthesis. And photosynthesis is the first and the foremost natural phenomenon for removing CO_2 from the atmosphere, and thus the first natural player to hold the climate in order.

Planting trees is a low-tech, most effective, most economic, and sustainable pathway to combat the climate menace of our times. Cutting down and/or minimizing carbon emissions by developing workable renewable-energy alternatives will complement the plantation mission for climate-change mitigation. It is estimated that there are some three trillion trees on Earth. Tom Crowther, a climate-change ecologist at the Swiss University ETH Zurich, says that plantation of an additional 1.2 trillion trees on Earth would accrue tremendous benefits in terms of absorbing atmospheric CO_2. There is enough space on Earth's land to accommodate an additional 1.2 trillion trees, according to Crowther.

What is the phenomenon that creates, enhances, sustains, and constantly evolves life on Earth and that gives the earth the status of a "Living Planet"? It is, of course, photosynthesis. All ecological processes on Earth are governed by photosynthesis. Chemosynthesis carried out by microorganisms in the pedosphere and aquatic environment too is complementary to photosynthesis. Chemosynthesis has also been referred to as "indirect" photosynthesis (Singh 2019) because the chemical energy in the compounds is owing to its origin in light. Boosting photosynthesis on Earth, therefore, is the key to solving all problems we encounter, including those of global warming and climate change. We can do that by planting as many trees on Earth's lands as possible.

A tree, according to the US-based Center for Urban Forest Research, sequesters 88 pounds (approximately 40 kg) of atmospheric CO_2 per year. Imagine how much the contribution would be for an extra 1.2 trillion trees on Earth when about three trillion trees on Earth are already put on this job!

Planting 1.2 trillion trees hints at a workable strategy putting the atmospheric CO_2 at bay. This strategy will accelerate the phenomenon of carbon sequestration in biomass and soils of the globe. To be realized, this phenomenon will need some time, until the plants attain maturity. Boosting photosynthesis to the extent of maximizing photosynthetic efficiency of the biosphere will help sequester atmospheric CO_2 to the extent that, in due course of time (in about 50 years, according to Tom Crowther), a balance in the global carbon cycle will be struck with a considerable proportion of the carbon bound in plant biomass and in soils.

There is ample space on Earth to accommodate more than 4 trillion trees. Where should the extra trees (1.2 trillion) be planted? Of course, amidst degraded forests, on abandoned lands, on field bunds, in agroforestry systems, in open spaces, on road sides, along railway tracts, in school compounds, in kitchen gardens, on riversides, and wherever we can find space suitable for planting trees.

We should not choose a single or just a few plant species to be planted. Rather, we must depend on a variety of tree species that can thrive in the climatic or ecological zone the plantation program is taken up. Native plants aimed at creating floral heterogeneity have to be an essential part of the strategy. Initially, the plants require adequate care, especially provision of water until the roots capture the depths from where they will suck water.

Since an anthropogenic increase of CO_2 concentration in the atmospheric pool is the major cause of global warming, the overall carbon cycle with accelerated movement from the atmosphere to another carbon pools is a real challenge in our times. At present, carbon concentrations in different pools are depicted in Figure 9.10. This picture of the current carbon cycle can be changed by enhancing photosynthesis, a phenomenon that can be driven by human beings by planting the maximum possible number of trees on Earth's surface. A phenomenal increase in photosynthetic efficiencies of Earth's ecosystems will bring a significant shift in carbon fluxes, accelerating the carbon movement from the atmospheric pool to plant biomass, soil, and litter pools. Halting deforestation, reduced

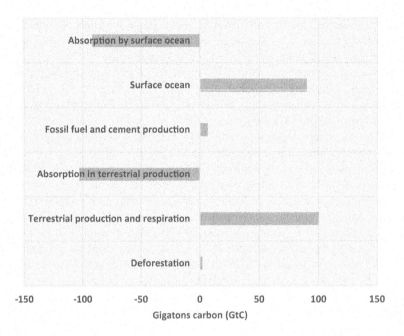

FIGURE 9.10 CO_2 fluxes (GtC) between Earth's atmosphere and other environmental components. About 3 GtC is added to the atmosphere annually. Earth's carbon reserves (GtC) are: surface carbon (1,020), deep ocean (38,000), fossil fuels (4,000), carbon rocks (65×10^6) and terrestrial ecosystems (2,070). (Modified from http://www.fao.org/3/ac836e/AC836E03.htm.)

emissions by means of fossil-fuel combustion, and ecologically sound food-production systems will ensure a substantial reduction in carbon fluxes from terrestrial ecosystems to the atmosphere.

An enhancement in photosynthetic efficiencies of Earth's ecosystems will also ensure an increase in carbon storages of surface ocean, deep ocean, and carbon rocks. Plantation of trees will ensure fortification of ecological regeneration in Earth's ecosystems leading to an enhancement in biodiversity and ecological integrity, which are an imperative for accelerating the processes leading to the attainment of sustainability.

CLIMATE-SMART AGRICULTURE

Our conventional agriculture, as we have observed elsewhere in this chapter, is also responsible for climate variation to a considerable extent. Since humanity has to inevitably depend on agriculture for nourishment, material progress, and sustenance, a sustainable future—which has to depend on a life-benevolent climate—will depend on the type of agriculture humanity develops and depends on. An ecologically sound and environmentally safe agriculture will eventually help humanity to usher in a sustainable future: a future to be witnessed by a healthy, vibrant, and happy humanity. The malady the modern agriculture suffers from is that it has been delinked from forests, rangelands/grasslands, natural soil health, livestock, etc. It is not a part of integral system as it used to happen in traditional agriculture. It is exclusively based on cultivated land and has linkages only with the external market. The first and the foremost strategy for an ecologically sound and environment-friendly agriculture is that it should be based on a system we call the farming system or agroecosystem, rather than just on croplands or cultivated lands.

A climate-smart agriculture should essentially be based on a carbon neutral (or carbon negative) system fulfilling all promises of a happy, healthy, and productive human life by providing all varieties of foods without disturbing or harming the environment. The development of a climate-smart agriculture is not only necessary but an imperative of our times.

WHY AN AGROECOSYSTEM APPROACH TO FOOD PRODUCTION?

An agroecosystem is an arbitrarily defined unit of nature with distinct energy flows and nutrient cycling using specific land use involving, for example, uncultivated land (forests, rangelands), cultivated land, and livestock woven into a complex unitary whole with organic linkages with each other, functionally oriented to produce foods and other life-supporting products, such as fodder, fiber, fuel, fertilizers, etc.—and performing vital ecological functions (Singh and Gaur 2007).

A typical agroecosystem, as operating in a traditional agriculture (e.g., in the Indian Himalayan region) comprises the components interrelated with each other through material and energy flows. An agroecosystem serves as a matrix (Figure 9.11), all the components nourishing each other through organic linkages that help maintain the dynamics of nutrient flows within an agroecosystem that, in turn, maintain ecosystem functionalities to be harnessed for the production of foods, fiber, and other commodities of human use (Figure 9.12).

Following are the major attributes of an agroecosystem approach to food production:

- Integrated multicomponent farming of different solar-powered ecosystems (operating under strong photosynthetic phenomenon) into a single functional unit
- Resource conservation-oriented development
- High degree of biodiversity (in-situ conservation of biodiversity in forests and rangelands)
- Increased photosynthetic efficiency and carbon sequestration: a contribution to maintaining a healthy climate
- Self-containment (production of essential agro inputs within the system itself)
- Shifting partial pressure for food, feed, and other life-support products on uncultivated areas
- Diversification of economic activities
- High resilience and sustainability of the agricultural system
- Food and livelihood security and food sovereignty

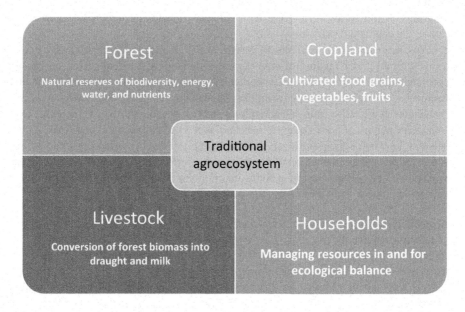

FIGURE 9.11 A traditional agroecosystem: a matrix of many components functioning in perfect synergy.

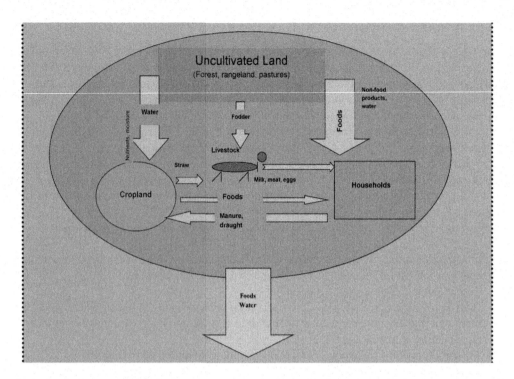

FIGURE 9.12 Organic linkages between agroecosystem components. Forest/rangeland is ecologically the most stable component nourishing the rest of the ecologically less stable ones. Biomass and energy flows within an agroecosystem are driven by photosynthesis.

AGROECOLOGY: THE PHILOSOPHY OF FOOD PRODUCTION

Agroecology is the holistic study of agroecosystems, including all ecological, environmental, and human elements. It is centered on the form, dynamics, and functions of their inter- and intrarelationships. It integrates agriculture with ecology. It helps us understand how agriculture revolves round the ecological principles.

Agroecology uses ecological principles to design, evolve, manage, and evaluate agroecosystems. It provides us with an understanding of vital biological processes, nutrient cycles, energy transformations, geographical and socioeconomic specificities, environmental and cultural specificities, and socioeconomic and cultural relationships in an integrated manner. Agroecology helps us evolve strategies for evolving what we call sustainable agriculture and ensures the future of agriculture. The strategies built on the principles of agroecology absorb and obey local cultures and traditions, such as food habits, festivities, and their ethical and aesthetic values. Applications of agroecology lead us to the protection, conservation, and augmentation of natural resources (forests, rangelands, livestock, fisheries, soil, water resources, etc.). It is an art of evolving, maintaining, and valuing overall farming cultures.

Agroecology embraces a number of its attributes, such as:

- Nutrient recycling (optimizing nutrient availability and balancing nutrient flow);
- Soil-fertility maintenance;
- Microclimate management;
- Species and genetic diversification in time and space; and
- Enhanced biological interactions, synergism, and augmentation of key ecological processes and services.

There are three principles of agroecology that can be operationalized in the development of ecologically sound agriculture (Singh et al. 2014):

1. Enhancing biodiversity complexity
2. Treating soil as an ecosystem
3. Maintaining cyclic flows of nutrients

Enhanced Biodiversity Complexity

There exists an intricate biodiversity–resilience–sustainability relationship in nature, which itself is a law of evolution in Earth's biosphere. The higher the level of biodiversity in an ecosystem (say, agroecosystem) the higher the degree of resilience. The higher the degree of resilience the higher the degree of sustainability. It is not just the number of species that matters in an ecosystem but the number of ecosystem functions that the species spark. Monocultures as common in the so-called Green-Revolution-type agriculture—as a rule—infuse vulnerability in an ecosystem. Biodiversity in various components of an agroecosystem (uncultivated lands/wild areas, croplands, and livestock, etc.) should be as high as possible at the species level as well as at their varieties (genotypes) levels subject to a specific geographical/agroecological region. Complexity is somewhat different, though it emanates from biodiversity. It implies intensive use of farm elements and diversification of ecosystem functionalities. A farm waste becomes an input in one or several functions, and eventually there is no waste or pollution on a farm.

Treating Soil as an Ecosystem

In implementing the agroecology principles, the soil is regarded as a living system as we have already discussed. Soil harbors an extremely diverse, abundant, and vibrant life, most of which is the microbial life. Fertility of the soil is directly proportional to the abundance and diversity of its life.

Maintaining Cyclic Flows of Nutrients

Forests in an agroecosystem comprise ecologically the most-stable ecosystem. The nutrient flow from a forest ecosystem to the ecologically more fragile cultivated land and recycling of nutrients from households (the largest consumers of the farm produce) and livestock to croplands will help an agricultural system to be more fertile, healthy, and productive.

The previously mentioned agroecological principles, when operationalized in agriculture, accrue many attributes as explained in Table 9.6. Agriculture based on operationalized agroecosystem principles will be distinctive from the conventional agriculture in several ways (Table 9.7). An ecosystem-based agriculture will be a sustainable alternative to conventional agriculture.

Organic agriculture is also a form of ecological agriculture in which cultivation practices are managed without the application of chemical fertilizers and pesticides. However, it does not call for an agroecosystem-based food production and ignores organic linkages among different components of an agroecosystems. Its main emphasis is on growing healthy foods. Agroecological processes also ensure production of healthy foods. Standards established for operationalizing organic agriculture, if clubbed with an agroecosystem-based farming, would further ameliorate the processes of a climate-smart agriculture. A climate-smart agriculture will assimilate and cultivate the values (Figure 9.13) worth ecological regeneration of our planet and a philosophy for a sane, ecologically just, sustainable, and happy living.

There can be no sustainable system without ecological sustainability of its own. Ecological sustainability, in essence, is woven around the phenomenon of photosynthesis. Sustainability, in fact, is rooted into an ecosystem fed by photosynthesis. An agroecosystem designed and developed on the basis of agroecological principles will be an ecologically sound, environmentally safe, climate-smart, and sustainable agricultural system.

TABLE 9.6
Attributes of Agroecological Strategies

Biodiversity Complexity	Living Soil (Soil Fertility Management)	Nutrient Flows
• High photosynthetic efficiency and increased carbon sequestration rates • Enhanced ecological integrity • Increased productivity • Soil and water conservation • Synergism • Conservation of beneficial insects • Intensive pollination • Enhanced resilience • Reduced risks of parasites, therefore, plant protection • Ultimate waste ends up as nutrient culture for the soil (no pollution) • More health per unit area (due to diverse foods accrued from biodiversity)	• Higher level of soil organic matter/soil organic carbon • Intensive food chains and food webs in soil • Increased N-fixation (due to proliferation of N-fixing bacteria) • Organic matter breakdown and increased mineralization (greater availability of nutrient ions for plant nutrition) • Facilitation of nutrient recycling • Healthy soil, healthy plants, healthy agroecosystem. • Enhanced photosynthetic efficiency of plants rooted in healthy soils • Soil functioning as a rich carbon pool	• Containing the nutrients within agroecosystem • Maintaining the health of fragile croplands • Amelioration of soil fertility • Prevention of pollution of water sources (streams, lakes, ponds) • Maintaining productivity of croplands using nutrients produced within the system • Reduced or no dependence on market for chemical fertilizers

TABLE 9.7
Some Contrasting Features of Conventional and Ecological Agriculture

Features	Conventional Agriculture	Ecological Agriculture
Farming system	Absent	Present
Soils	Poor	Healthy
Fragility	High	Low
Diversity/heterogeneity	Minimum	High or extreme
Niche/comparative advantage	Poor	Rich
Adaptation capability	Poor	High
Complexity	Less	High to extreme
Vulnerability	High	Low
Resilience	Low	High
Water-use efficiency	Poor	High
Inputs	External	Internal
Photosynthetic efficiency	Poor	High
Carbon emission	High	Minimum
Carbon sequestration	Low	Potentially high
Productivity	Moderate	High
Response to climate	Negative	Positive
Sustainability	Poor	High

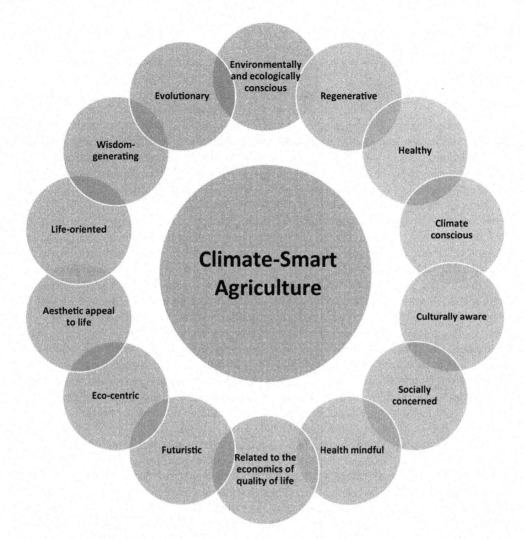

FIGURE 9.13   Climate-benevolent attributes of a climate-smart agriculture.

VEGETARIANISM: THE PHILOSOPHY OF EARTH-CARING NUTRITION

We have documented elsewhere the livestock sector contribution to GHG emissions. Most of the GHG contributions are thanks to intensive animal farming for meat production. The meat industry itself is notorious and a great burden on the entire biosphere, threatening the lives of millions of people worldwide, apart from extinction of so many species due to the gradually intensifying climate change. A recent report compiled by Climate Nexus (2016) reads:

> If people continue to eat meat at current rates, half of the carbon budget would be lost in three decades. By 2050, according to the modeling, switching to diets of limited meat consumption can cut emissions by a third while saving 5 million lives, vegetarian diets could reduce emissions by 63 percent and save 7 million lives, and vegan diets could reduce emissions by 70 percent and save 8 million lives...The lowest level of meat consumption—widespread adoption of the vegan diet—could avoid more than 8 million deaths by 2050. A vegetarian diet would save 7.3 million lives.

Meat production is not just confined to excessive GHG emissions. It is now being regarded as a leading cause of mass extinction. Gabrielle (2015) has predicted that in the next 35 years or so, as many as 37 percent of the world's species will go extinct. Meat production, on a global scale, is an environmental disaster, and by 2050 it will be a complete and unthinkable disaster (Mellino 2016). The vegetarian food habits are not just non-violent or a compassionate lifestyle, but can also help save the planet.

Vegetarianism is not just a food habit. It is more than that: it is a philosophy of Earth-caring nutrition. All those organisms that depend on vegetation for their nutrition stay closest to the abundance of "living" energy being generated through photosynthesis. Populations of the herbivores or vegans are just one step away from photosynthesis and, therefore, the most stable ones in the Kingdom Animalia. Less-stable populations are of first-order carnivores and still less are those of second-order carnivores, the least stable being those of the top carnivores. In other words, the most sustainable life prospers nearest to photosynthesis: first plants and then herbivores. Let us understand this with the following example (Figure 9.14).

Suppose photosynthesizers (the producers) assimilate 100 cal of energy of the food, only 10 cal are consumed by the herbivores, and only a hundredth fraction of a calorie passes on to the top carnivores at the farthest end of photosynthesis. As per the oft-quoted Ten Percent Law, as introduced by Lindeman (1942), only 10% energy is transferred from a trophic level to a the next higher trophic level in the form of organic food. Thus, the trophic level farthest from photosynthesis is left with the least amount of energy for consumption. This is the reason that only a few trophic levels (some five or six) exist in nature because the food synthesized by photosynthesis cannot nourish endless trophic numbers. Carnivory among animals is very expensive for the biosphere. The biosphere, in fact, has been affording carnivory sustainably as wild animals help compensate the same by their contribution to ecological regeneration and striking an ecobalance. However, the prevailing meat-eating habits of the human species are increasingly becoming unaffordable for the biosphere to sustain. Human activities are uncontrollable, and human greed is insatiable to the extent that they have gradually assumed an ecocidal proportion from which emanates the life-annihilating climate crisis.

Laws of the biosphere also apply to the human race. The energy-flow pattern in an ecosystem also applies to a human society. From the ecological point of view, we can imagine how exorbitantly expensive the meat diet is for the human race! Being vegetarian means flourishing with the nature's foundation of plentiful energy—the Kingdom Plantae. Vegetarian foods can nourish, support, and sustain much larger populations than those of the meat eaters. Being vegetarian means living closest to the phenomenon of photosynthesis. Deviation away from photosynthesis means losing access to a plentiful availability of energy through foods. Living closest to photosynthesis, that is, being vegetarian, means prevailing with the highest degree of sustainability (Singh 2019). Vegetarianism is not just about a food habit; it is a living philosophy so vital for healing climate disruption.

The climate itself is a function of a complex mix of operating environmental factors. Human interactions with natural resources leading to changes in structure, composition, and functioning of the biosphere's ecosystems and changes in the composition of Earth's atmosphere are phenomenally affecting the climate system. Development of an intervention involving photosynthesis-boosting botanical strategies, as previously discussed, can be vital in putting the earth's climate system back into order. The biosphere's and its living organisms' health, vibrancy, functioning, and sustainability primarily depend on the climate pattern. A climate-smart planet, in essence, is a sustainable planet.

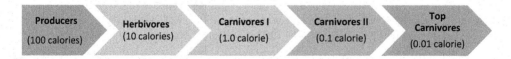

FIGURE 9.14 Energy flow through the food chain in an ecosystem.

SUMMARY

Climate change is the grimmest reality of our times—a reality that is enlarging itself dynamically and becoming more and more formidable year after year. The dire fact facing the contemporary times is that our Living Planet is now stalked in an Age of Climate Change. Solar output, Earth–Sun geometry, stellar dust, continental drift, volcanic activity, mountain building, oceanic heat exchange, atmospheric chemistry, atmospheric albedo, and surface albedo have been the major interacting factors building up a kind of climate that life on Earth has evolved and perpetuated in. However, the unabated changes in the climate pattern are largely attributable to anthropogenic factors. In fact, the planet now is stalked in the midst of an enhanced greenhouse effect. Planet Earth is potent enough to cope with and recuperate the losses its ecosystems suffer from by means of self-regulatory processes. Indiscriminate and ever-intensifying anthropogenic activities, however, are gradually overpowering the planet's inherent potencies of ecological regeneration. There are some nine dimensions of environmental disruptions, viz., land-use change, biodiversity erosion, nitrogen and phosphorus cycles, global freshwater use, ocean acidification, chemical changes, atmospheric-aerosol loading, stratospheric-O_3 depletion, and climate change.

Climate change and the consequent altering temperature patterns are inducing microevolutionary changes (disruptions of long-standing patterns of natural selection on plant physiological traits), macroevolutionary changes (changes in plant physiology and species diversification rates following past climatic change events), and eco-evolutionary changes (limitations in plant physiological responses to current and future climate changes and direct effect on fitness and survival). Evolutionary responses to climate change at the physiological level happen to be a key challenge facing plant physiologists and climate experts. A deeper understanding of the potential of the evolutionary responses would help in predicting plant responses to climate change more precisely. The physiological pathways we encounter in the plants are an outcome of long-term changes in the environment spanning millions of years.

Climate change can bring substantial alterations in mutualism, competition, and antagonisms among plant species in a community and drive the evolution of their physiology. Climate change has multiple dimensions and triggering effects, and it affects every phenomenon of the biosphere, including every aspect of living organisms' physiologies. Alteration in plant physiology in response to climate change is the root cause of all ecological changes taking place—most likely irreversible.

Climate change at a global scale is all set to exert its phenomenal impact on the structure, composition, distribution, functions, and ecological dynamics of the earth's forest ecosystems. Agriculture is both a driver and a victim of climate change. Agricultural systems are more vulnerable to climate change than the natural ecosystems. With further soaring of temperatures, weather cycles will become more erratic and drought–flood cycles will be more damaging to agriculture.

Independent of other variables due to climate change, the steadily increasing atmospheric CO_2 level would have a profound direct effect on the physiology, biochemistry, and growth of food crops. C_3 plants, which account for 95% of all the plants, would respond more conspicuously to elevated CO_2 levels than C_4 plants. Unabated climate change may also destroy crop biodiversity. Many other indirect climate-change impacts will include undesirable alterations in agroecosystems. Changes in the soil microflora, microfauna, mesofauna, and macrofauna, and in vegetation composition, would invite biotic stresses, such as competitions due to proliferation of weeds and suitable conditions for invasion by exotic plants. There can be a spurt in the populations of pests and pathogens (including expansion of ranges and seasons) that may cause large-scale crop damages. Animal husbandry suffers far more from climate change than crop husbandry.

Strategies for coping with climate change must include preparedness, adaptation, and mitigation. A climate-smart planet is a sustainable planet. Our workable strategies to build up a climate-smart planet must focus on the realization of (1) ecological integrity and ecological security; (2) living soil to nurture roots of life; and (3) forests for a healthy, vibrant, and sustainable planet.

Development of climate-smart agricultural systems is a part of the strategy for the development of a climate-smart planet. The first and the foremost strategy for an ecologically sound and environment friendly agriculture is that it should be based on a system we call the farming system or agroecosystem, rather than just on croplands or cultivated lands. A climate-smart agriculture should essentially be based on a carbon-neutral (or carbon-negative) system fulfilling all promises of a happy, healthy, and productive human life by providing all varieties of foods without disturbing or harming the environment. A climate-smart agriculture should be developed on the principles of agroecology, which involve biodiversity complexity, living soil, and cyclic flows of nutrients. A climate-smart agricultural system is environmentally and ecologically conscious, climate conscious, regenerative, healthy, life-oriented, wisdom generating, eco-centric, futuristic, related to the economics of the quality of life, socially concerned, health mindful, full of aesthetic appeal, and evolutionary.

REFERENCES

Alberto, F. J., Aitken, S. N., Alia, R., Gonzalez-Martinez, S. C., Hanninen, H., Kremer, A., Lefevre, F. Lenormand, T., Yeaman, S. and Whetten, R. 2013. Potential for evolutionary responses to climate change: Evidence from tree populations. *Glob. Change Biol.* 19: 1645–1661.

Alexander, J. M., Diez, J. M. and Levine, J. M. 2015. Novel competitors shape species' responses to climate change. *Nature* 525: 515–518.

Anderson, J. T. 2016. Plant fitness in a rapidly changing world. *New Phytol.* 210: 81–87.

Anderson, J. T., Inouye, D. W., McKinney, A. M., Colautti, R. I. and Mitchell-Olds, T. 2012. Phenotypic plasticity and adaptive evolution contribute to advancing flowering phenology in response to climate change. *Proc. Biol. Sci.* 279: 3843–3852.

Arakaki, M., Christin, P. A., Nyffeler, R., Lendel, A., Eggli, U., Ogburn, R. M., Spriggs, E., Moore, M. J. and Edwards, E. J. 2011. Contemporaneous and recent radiations of the world's major succulent plant lineages. *Proc. Natl. Acad. Sci. USA* 108: 8379–8384.

Bale, J., Masters, G. J., Hodkins, I. D., Awmack, C., Bezemer, T. M., Brown, V. K., Buterfield, J., et al. 2002. Herbivory in global climate change research: Direct effects of rising temperature on insect herbivores. *Global Change Biol.* 8: 1–16.

Bauer, I. E., Apps, M. J., Bhatti, J. S. and Lal, R. 2006. Climate change and terrestrial ecosystem management: Knowledge gaps and research needs. In: Bhatti, J. S., Lal, R., Apps, M. and Price, M. (eds.) *Climate Change and Managed Ecosystems.* Boca Raton, FL: Taylor & Francis Group, CRC Press, p. 411.

Becklin, K. M., Anderson, J. T. Gerhart, L. M., Wadgymar, S. M., Wessinger, C. A. and Ward, J. K. 2016. Examining plant physiological responses to climate change through an evolutionary lens. *Plant Physiol.* 172: 635–649. doi: https://doi.org/10.1104/pp.16.00793.

Berner, R. 2006. Geocarbsulf: A combined model for Phanerozoic atmospheric O_2 and CO_2. *Geochim. Cosmochim. Acta* 70: 5653–5664.

Bone, R. E., Smith, J. A., Arrigo, N. and Buerki, S. 2015. A macro-ecological perspective on Crassulacean acid metabolism (CAM) photosynthesis evolution in Afro-Madagascan drylands: Eulophiine orchids as a case study. *New Phytol.* 208: 469–481.

Christin, P. A., Osborne, C. P., Chatelet, D. S., Columbus, J. T., Besnard, G., Hodkinson, T. R., Garrison, L. M. and Vorontsova, M. S. 2013. Anatomical enablers and the evolution of C4 photosynthesis in grasses. *Proc. Natl. Acad. Sci. USA.* 110: 1381–1386.

Climate Nexus. 2016. Eating less meat could save 5 million lives, cut carbon emissions by 33%. http://ecowatch.com/2016/03/22/eat-less-meat-save-lives-cut-carbon (Retrieved 22 March 2016).

Currano, E. D., Wilf, P., Wing, S. L., Labandeira, C. C., Lovelock, E. C. and Royer, D. L. 2008. Sharply increased insect herbivory during the Paleocene-eocene thermal maximum. *Proc. Natl. Acad. Sci.* 105: 1960–1964.

Das, H. P. 2004. Adaptation strategies required to reduce vulnerability in agriculture and forestry to climate change, climate variability and climate extremes. In: *World Meteorological Organization (WMO): Management Strategies in Agriculture and Forestry for Mitigation of Greenhouse Gas Emissions and Adaptation to Climate Variability and Climate Change.* Report of CAgM Working Group. Technical Note No. 202, WMO No. 969. Geneva, Switzerland: WMO, pp. 41–92.

Edwards, E. J. and Ogbum. R. M. 2012. Angiosperm responses to a low-CO_2 world: CAM and C_4 photosynthesis as parallel evolutionary trajectories. *Int. J. Plant Sci.* 173: 724–733.

Edwards, E. J. and Smith, S. A. 2010. Phylogenetic analyses reveal the shady history of C_4 grasses. *Proc. Natl. Acad. Sci. USA* 107: 2532–2537.

Edwards, E. J., Osborne, C. P., Stromberg, C. A. E., Smith, S. A., Bond, W. J., Christin, P. A., Cousins, A. B., Duvall, M. R., Fox, D. L. and Freckleton, R. P. 2010. The origins of C4 grasslands integrating evolutionary and ecosystem science. *Science* 328: 587–591.

Ehleringer, J. R. and Monson, R. K. 1993. Evolutionary and ecological aspects of photosynthetic pathway variation. *Annu. Rev. Ecol. Syst.* 24: 411–439.

Ehrenfeld, D. 2005. The environmental limits to globalization. *Conserv. Biol.* 19 (2): 318–326.

FAO, IFAD and WFP. 2015. *The State of Food Insecurity in the World: Meeting the 2015 International Hunger Targets—Taking Stock of Uneven Progress.* Rome, Italy: FAO. 56 p.

Gabrielle, C. 2015. Your meat-Eating habit is killing more than cows. http://www.motherjones.com/bluemarble/2015/08/your-meat-eating-habit-killing-more-just-cows (Retrieved August 21, 2015).

Garrett, K. A., Dendy, S. P., Frank, E. E., Rouse, M. N. and Travers, S. E. 2006. Climate change effects on plant disease: Genomes to ecosystems. *Annu. Rev. Phytopathol.* 44: 489–509.

Goodland, R. and Anhang, J. November–December 2009. *Livestock and Climate Change: What if the Key Actors in Climate Change were Pigs, Chickens and Cows?* Washington, DC: Worldwatch Institute, pp. 10–19.

GRAIN. 2011. Food and climate change: The forgotten link. *Seedling*, September 2011: 1–5.

Hargesheimer, K. 2007. Organic, No-till Farming. Gardens/ Mini-Farms Network. http://time.com/4266874/vegetarian-diet-climate-change (Retrieved March 22, 2016).

IPCC. Climate Change 2007. *The Physical Science Basis. Contribution of Working Group I to the Fourth Assessment Report of the Intergovernmental Panel on Climate Change.* Cambridge, UK: Cambridge University Press.

Karnosky, D. F. Pregitzer, K. S., Zak, D. R. Kubiske, M. E. Hendrey, G. R., Weinstein, D., Nosal, M. and Percy, K. E. 2005. Scaling ozone responses of forest trees to the ecosystem level in a changing climate. *Plant, Cell Environ.* 28: 965–981.

Karnosky, D. F., Percy, K. E., Xiang, B., Callan, B., Noormets, A., Mankovska, B., Hopkin, A. et al. 2002. Interacting elevated CO_2 and tropospheric O_3 predisposes aspen (*Populus tremuloides* Michx.) to infection by rust (*Melampsora medusa f.sp. tremuloidae*). *Glob. Change Biol.* 8: 329–338.

Keuskamp, D. H., Sasidharan, R. and Pierik, R. 2010. Physiological regulation and functional significance of shade avoidance responses to neighbours. *Plant Signal Behav.* 5: 655–662.

Lau, J. A., Shaw, R. G., Reich, P. B. and Tiffin, P. 2014. Indirect effects drive evolutionary responses to global change. *New Phytol.* 201: 335–343.

Lenoir, J., Gegout, J. C., Marquet, P. A., de Rufftray, P. and Brisse, H. 2008. A significant upward shift in plant species optimum elevation during the 20th century. *Science* 320: 1768–1771.

Lindeman, R.L. 1942. The trophic-dynamic aspect of ecology. *Ecology* 23: 399–418.

Loarie, S. R., Duffy, P. B., Hamilton, H., Asner, G. P., Field, C. B. and Ackerly, D. D. 2009. The velocity of climate change. *Nature* 462: 1052–1055.

Loria, K. 2018. Earth just hit a terrifying milestone for the first time in more than 800,000 years. Business Insider India, May 08, 2018 (https://www.businessinsider.in/Earth-just-hit-a-terrifying-milestone-for-the-first-time-in-more-than-800000-years/articleshow/64080483.cms).

Lu, Y. and Conrad, R. 2005. *In situ* stable isotope probing of methanogenic archaea in the rice rhizosphere. *Science* 309: 1088–1090.

Mackenzie, F. T. 2003. *Our Changing Planet: An Introduction to Earth System Science and Global Environmental Change*, 3rd edn. Upper Saddle River, NJ: Prentice Hall. 580 p.

Medvigy, D. and Beaulieu, C. 2012. Trends in daily solar radiation and precipitation coefficient of variation since 1984. *J. Clim.* 25: 1230–339.

Mellino, C. 2016. The shocking consequences of the world's meat addiction. *Ecowatch.* http://ecowatch.com/2016/03/04/world-meat-addiction (Retrieved March 4, 2016).

Mitchell-Olds, T., Willis, J. H. and Goldstein, D. B. 2007. Which evolutionary processes influence natural genetic variation for phenotypic traits? *Nat. Rev. Genet.* 8: 845–856.

Moore B. A. and Allard G. B. 2008. Climate change impacts on forest health. Forest Health & Biosecurity Working Papers FBS/34E. Forest Resources Development Service, Forest Management Division. Rome: FAO.

Norby, R. J., De Lucia, E. H., Gielen, B., Calfapietra, C., Giardina, C. P., King, J. S., Ledford, J. et al. 2005. Forest response to elevated CO_2 is conserved across a broad range of productivity. *PNAS* 102: 18052–18056.

Norby, R. J., Rustad, L. E., Dukes, J. S., Ojima, D. S., Parton, W. J., Del Grosso, S. J., McMurtrie, R. E. and Pepper, D. A. 2007. Ecosystem responses to warming and interacting global change factors. In: Canadell, J. G., Pataki, D. and Pitelka, L. eds. *Terrestrial Ecosystems in a Changing World.* The IGBP Series, Berlin, Germany: Springer-Verlag. pp. 23–36.

Parmesan, C. 2006. Ecological and evolutionary responses to recent climate change. *Annu. Rev. Ecol. Evol. Syst.* 37: 637–669.

Patz, J. A., Campbell-Lendrum, D., Holloway, T. and Foley, J. A. 2005. Impact of regional climate change on human health. *Nature.* 438: 310–317.

Perrone, T. 2018. How agriculture and climate change are related: Causes and effects. https://www.lifegate.com/people/news/agriculture-and-climate-change-causes-effects-impacts (Retrieved April 29, 2019).

Sage, R. F., Christin, P-A. and Edwards, E. J. 2011. The C4 plant lineages of planet Earth. *J. Exp. Bot.* 62: 3155–3169.

Sage, R. F., Sage, T. L. and Kocacinar, F. 2012. Photorespiration and the evolution of C4 photosynthesis. *Annu. Rev. Plant Biol.* 63: 19–47.

Scheehle, E. 2002. Emissions and projections of non CO_2 greenhouse gases from developing countries, 1990–2020. http://www.epa.gov/ttn/chief/conference/ei10/ghg/scheehle.pdf. p.73.

Shaw, R. G. and Etterson, J. R. 2012. Rapid climate change and the rate of adaptation: Insight from experimental quantitative genetics. *New Phytol.* 195: 752–765.

Singh, V. 2018. *An Analysis of Mountain Agro-Ecosystems.* Beau Bassin, Mauritius: Lambert Academic Publishing. 64 p.

Singh, V. 2019. *Fertilizing the Universe: A New Chapter of Unfolding Evolution.* London, UK: Cambridge Scholars Publishing. 286 p.

Singh, V. and Gaur, R. D. 2007. Mountain Agriculture in the Indian Himalaya: Specificities, Scenarios, Sustainability and Strategies. In: Rawat, M. S. S., Singh, M. and Singh, V. (eds.) *Management Strategies for the Indian Himalaya: Development and Conservation.* Srinagar, Garhwal: Department of Geography, HNB Garhwal University, pp. 79–99.

Singh, V., Nautiyal, N., Apparusu, S. K. and Rawat, M. S. S. (eds.). 2011. *Climate Change in the Himalayas: Preserving the Third Pole for Cooling the Earth.* New Delhi, India: Indus Publishing Company. 302 p.

Singh, V., Rastogi, A., Nautiyal, N. and Negi, V. 2017. Livestock and climate change: The key actors and the sufferers of global warming. *Indian J. Anim. Sci.* 87 (1): 11–20.

Singh, V., Shiva, V. and Bhatt, V. K. 2014. *Agroecology: Principles and Operationalisation of Sustainable Mountain Agriculture.* New Delhi, India: Navdanya. 64p+viii.

Sirohi, S. and Michaelowa, A. 2007. Sufferer and cause: Indian livestock and climate change. *Clim. Change* 85: 285–298.

Skolimowski, H. 1991. *Dancing Shiva in the Ecological Age.* New Delhi, India: Clarion Books. 182 p.

Spriggs, E. L., Christin, P. A. and Edwards, E. J. 2014. C_4 photosynthesis promoted species diversification during the Miocene grassland expansion. *PLOS ONE* 9 (8): e105923.

Springer, C. J., Orozco, R. A., Kelly, J. K. and Ward, J. K. 2008. Elevated CO_2 influences the expression of floral-initiation genes in Arabidopsis thaliana. *New Phytol.* 178: 63–67.

Steinfeld, H., Gerber, P., Wassennar, T. D., Caster, V. and Haan, C. 2006. *Livestock's Long Shadow: Environmental Issues and Options.* Rome, Italy: FAO.

Stone, J. M. R., Bhatti, J. S. and Lal, R. 2006. Impacts of climate change on agriculture, forest and wetland ecosystems: Synthesis and summary. In: Bhatti, J. S., Lal, R. Apps, M. and Price, M. (eds.). *Climate Change and Managed Ecosystems.* Boca Raton, FL: Taylor & Francis Group, CRC Press, pp. 399–409.

Thomson, M., Gamage, D., Hirotsu, N., Martin, A. and Seneweera, S. 2017. Effects of elevated carbon dioxide on photosynthesis and carbon partitioning: A perspective on root sugar sensing and hormonal crosstalk. *Front Physiol.* 8: 578. doi: 10.3389/fphys.2017.00578. eCollection 2017.

Tripati, A. K., Roberts, C. D. and Eagle, R. A. 2009. Coupling on CO_2 and ice sheet stability over major climate transitions of the late 20 million years. *Science* 326: 1394–1397.

Trumbore S 2006. Carbon respired by terrestrial ecosystems—Recent progress and challenges. *Global Change Biol.* 12: 141–153.

Wahal, V., Ponnu, J., Schlereth, A., Arrivault, S., Langenecker, T., Franke, A., Feil, R., Lunn, J. E., Stitt, M. and Schmid, M. 2013. Regulation of flowering by trehalose-6-phosphate signaling in *Arabidopsis thaliana.* *Science* 339: 704–707.

Ward, J. K. and Kelly, J. 2004. Scaling up evolutionary responses to elevated CO_2: Lessons from Arabidopsis. *Eco. Lett.* 7: 427–440.

Williams, J. 2004. *Sustainable Development on Africa: Is the Climate Right?* IRI Technical Report IRI-TR/05/1. New York: The International Research Institute for Climate Prediction, Palisades; *Clim. Change* 43: 651–681.

WEBSITES

http://time.com/4266874/vegetarian-diet-climate-change/
http://www.fao.org/3/ac836e/AC836E03.htm
http://www.finestquotes.com/select_quote-category-Agriculture-page-0.htm
http://www.plantphysiol.org/content/160/4/1675
http://www.plantphysiol.org/content/172/2/635
https://earthobservatory.nasa.gov/features/Arrhenius
https://earthobservatory.nasa.gov/features/Tyndal
https://earthobservatory.nasa.gov/world-of-change/DecadalTemp
https://www.freeimages.com/photo/banyan-tree-1516030
https://www.google.com/search?q=The+Great+Banyan+Tree
https://www.grain.org/article/entries/4357-food-and-climate-change-the-forgotten-link
https://www.grain.org/en/category/539
https://www.lifegate.com/people/news/agriculture-and-climate-change-causes-effects-impacts
https://www.ncbi.nlm.nih.gov/pmc/articles/PMC5047093/
https://www.sciencedirect.com/science/article/pii/S0963996909003421

Index

Note: Page numbers in italic and bold refer to figures and tables, respectively.